W9-BMA-045

MODULAR SERIES
ON SOLID STATE DEVICES

VOLUME IV
Field Effect Devices

MODULAR SERIES
ON SOLID STATE DEVICES
Robert F. Pierret and Gerold W. Neudeck, Editors

VOLUME IV
Field Effect Devices

ROBERT F. PIERRET
Purdue University

ADDISON-WESLEY PUBLISHING COMPANY
READING, MASSACHUSETTS
MENLO PARK, CALIFORNIA
LONDON • AMSTERDAM
DON MILLS, ONTARIO
SYDNEY

This book is in the
Addison-Wesley Modular Series on Solid State Devices

Library of Congress Cataloging in Publication Data

Pierret, Robert F.
Field effect devices.

(Modular series on solid state devices ; v. 4)
"EE-305."
Bibliography: p.
Includes index.
1. Field-effect transistors. 2. Metal oxide semiconductors.
3. Metal oxide semiconductor—field-effect transistor. I. Title.
II. Series: Pierret, Robert F. Modular series on solid state
devices ; v. 4.
TK 7871.95.P53   621.3815'284   81-15035
ISBN 0-201-05323-3   AACR2

ISBN 0-201-05323-3
ABCDEFGHIJ-AL-898765432

Foreword

Solid state devices have attained a level of sophistication and economic importance far beyond the highest expectations of their inventors. The bipolar and field-effect transistors have virtually made possible the computer industry which in turn has created a completely new consumer market. By continually offering better performing devices at lower cost per unit, the electronics industry has penetrated markets never before addressed. One fundamental reason for such phenomenal growth is the enhanced understanding of basic solid state device physics by the modern electronics designer. Future trends in electronic systems indicate that the digital system, circuit, and IC layout design functions are being merged into one. To meet such present and future needs we have written this series of books aimed at a qualitative and quantitative understanding of the most important solid state devices.

Volumes I through IV are written for a junior, senior, or possibly first-year graduate student who has had a reasonably good background in electric field theory. With some deletions these volumes have been used in a one semester, three credit-hour, junior-senior level course in electrical engineering at Purdue University. Following this course are two integrated-circuit-design and two IC laboratory courses. Each volume is written to be covered in 12 to 15 fifty-minute lectures.

The individual volumes make the series useful for adoption in standard and non-standard format courses, such as minicourses, television, short courses, and adult continuing education. Each volume is relatively independent of the others, with certain necessary formulas repeated and referenced between volumes. This flexibility enables one to use the series continuously or in selected parts, either as a complete course or as an introduction to other subjects. We also hope that students, practicing engineers, and scientists will find the volumes useful for individual instruction, whether it is for reference, review, or home study.

A number of the standard texts on devices have been written like encyclopedias, packed with information, with little thought to how the student learns or reasons. Texts that are encyclopedic in nature are difficult for students to read and are often barriers to their understanding. By breaking the material into smaller units of information and by writing for students (rather than for our colleagues) we hope to enhance their understanding of solid state devices. A secondary pedagogical strategy is to strike a healthy balance between the device physics and practical device information.

The problems at the end of each chapter are important to understanding the concepts presented. Many problems are extensions of the theory or are designed to reinforce particularly important topics. Some numerical problems are included to give the reader an intuitive feel for the size of typical parameters. Then, when approximations are stated or assumed, the student will have confidence that certain quantities are indeed orders of magnitude smaller than others. The problems have a range of difficulty, from very simple to quite challenging. We have also included discussion questions so that the reader is forced into qualitative as well as quantitative analyses of device physics.

Problems, along with answers, at the end of the first three volumes represent typical test questions and are meant to be used as review and self-testing. Many of these are discussion, sketch, or "explain why" types of questions where the student is expected to relate concepts and synthesize ideas.

We feel that these volumes present the basic device physics necessary for understanding many of the important solid state devices in present use. In addition, the basic device concepts will assist the reader in learning about the many exotic structures presently in research laboratories that will likely become commonplace in the future.

West Lafayette, Indiana

R. F. Pierret
G. W. Neudeck

Contents

Introduction 1

1 The Junction Field Effect Transistor

1.1 Introduction 3
1.2 Qualitative Theory of Operation 5
1.3 Quantitative $I_D - V_D$ Relationships 10
1.4 ac Response 16
Problems 18

2 Ideal MOS Device Statics

2.1 The Ideal MOS Structure 21
2.2 Visualization of the Static State 22
2.2.1 The Energy Band Diagram 22
2.2.2 The Static State 25
2.3 Semiconductor Electrostatics 28
2.3.1 Definition of Parameters 28
2.3.2 Exact Solution 30
2.3.3 Delta-depletion Solution 32
2.4 Gate Voltage Relationship 36
2.5 Summary and Concluding Comments 39
Problems 40

3 Capacitance – Voltage Characteristics

3.1 Qualitative Theory 43
3.2 Delta-depletion Analysis 48

3.3 Exact Charge Analysis 50
3.4 Practical Considerations/Deep Depletion 53
Problems 56

4 Deviations from the MOS Ideal

4.1 Metal-semiconductor Workfunction Difference 59
4.2 Mobile Ions in the Oxide 64
4.3 The Fixed Oxide Charge 69
4.4 Interfacial Traps 70
4.5 Summary and Concluding Comments 77
Problems 77

5 MOS Field Effect Transistors

5.1 Qualitative Theory of Operation 81
5.2 Quantitative $I_D - V_D$ Relationships 85
5.2.1 The Effective Mobility 85
5.2.2 General Analysis 87
5.2.3 Square-law Theory 89
5.2.4 Bulk-charge Theory 90
5.3 Threshold Considerations 93
5.3.1 Threshold Voltage Relationships 93
5.3.2 Threshold, Terminology, and Technology 94
5.3.3 Threshold Adjustment 95
5.3.4 Back Biasing 97
5.4 ac Response 98
5.4.1 Small Signal Equivalent Circuits 98
5.4.2 Cutoff Frequency 100
5.4.3 Small Signal Characteristics 100
5.5 Summary and Concluding Comments 102
Problems 103

Suggested Readings 107

Appendix

List of Symbols 109

Index 113

Introduction

This volume is intended to be an introduction to field-effect devices. Three of the more basic members of the field-effect-device family are examined in some detail to expose the reader to relevant terms, concepts, models, and analytical procedures. Chapter 1 is devoted to the Junction Field Effect Transistor (J-FET). If the reader is familiar with *pn*-junction operation, the J-FET provides a conceptual bridge between the *pn*-junction devices and the "pure" field-effect devices considered later in the volume. The J-FET discussion, moreover, serves as a first run through on terminology, analytical procedures, etc., thereby making the subsequent discussion easier to present and understand. Chapters 2 through 4 deal with the simplest of the Metal–Oxide–Semiconductor (MOS) structures, namely, the MOS-Capacitor. Because the MOS-C functions both as a diagnostic tool and forms the heart of more complex MOS devices, the effort expended to characterize the MOS-C is more than worthwhile. Chapter 2 explores the description of the ideal MOS-C in the static state, Chapter 3 covers the MOS-C capacitance–voltage characteristics, and Chapter 4 describes common deviations from the ideal in MOS structures. The culmination of the volume, Chapter 5, treats the commercial giant of the group, the Metal–Oxide–Semiconductor Field-Effect Transistor (MOSFET). The MOSFET presentation encompasses the operation and analysis of the basic transistor configuration.

It is expected that the reader will find the presentation in this volume to be at a somewhat more advanced level than Volumes I through III. The stepup in level is dictated for the most part by the nature of the subject matter. However, having mastered the concepts contained in Volumes I through III, the reader should have little difficulty in handling a slightly more challenging presentation. If desired, the material coverage may be simplified without a serious loss in continuity by omitting the exact solution portion of Section 2.3, Section 3.3, and the bulk-charge theory discussed in Subsection 5.2.4.

1 / The Junction Field Effect Transistor

1.1 INTRODUCTION

Students in electrical engineering are sometimes surprised to learn that, historically, the field-effect phenomenon was the basis for the first type of solid state transistor ever proposed. Indeed, field-effect transistors predate the bipolar junction transistor by approximately 20 years! As recorded in a series of patents filed in the 1920s and 1930s, J. E. Lilienfeld in the United States and O. Heil working in Germany independently conceived and investigated the transistor shown in Fig. 1.1. The device worked on the principle that a voltage applied to the metallic plate modulated the conductance of the underlying semiconductor, which in turn modulated the current flowing between ohmic contacts A and B. This phenomenon, where the conductivity of a semiconductor is modulated by an electric field applied normal to the surface of the semiconductor, has been subsequently named the *field effect*.

The early field-effect transistor proposals were of course far ahead of their time. Modern-day semiconductor materials were just not available and technological imma-

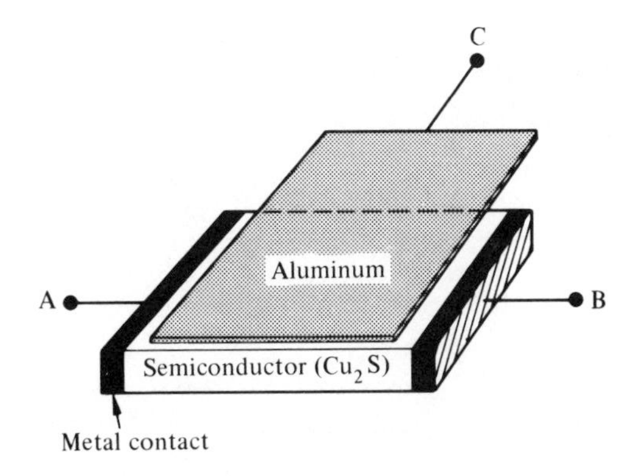

Fig. 1.1 Idealization of the Lilienfeld transistor.

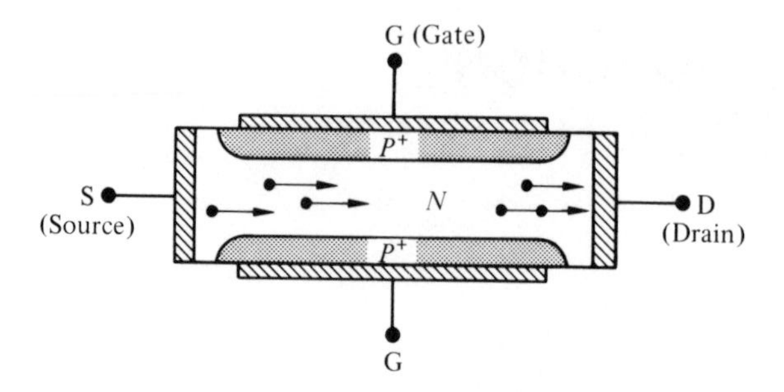

Fig. 1.2 Schematic cross-sectional diagram of the Junction Field Effect Transistor.

turity, in general, simply forced the field-effect structure to lie dormant for many years. With the development of other solid state devices in the late 1940s, however, the field-effect concept was revived. The first modern-day field-effect device, the Junction Field Effect Transistor (J-FET), was proposed by W. Shockley in 1952. In the J-FET, pictured schematically in Fig. 1.2, *p-n* junctions were used to modulate the semiconductor conductivity instead of a metallic plate, the A and B contacts became known as the source and drain, and the field-effect electrode was named the gate.

Originally the Fig. 1.2 device was named the Unipolar Transistor to distinguish this transistor from the Bipolar Junction Transistor and to emphasize that *only one type of carrier was involved in the operation* of the new device. For the structure pictured in Fig. 1.2, normal operation of the transistor can be described totally in terms of the electrons flowing in the *n*-region from the source to the drain. The source (S) terminal gets its name from the fact that the carriers contributing to the current flow move from the external circuit into the semiconductor at this electrode. The carriers leave the semiconductor, or are "drained" from the semiconductor, at the drain (D) electrode. The gate gets its name from

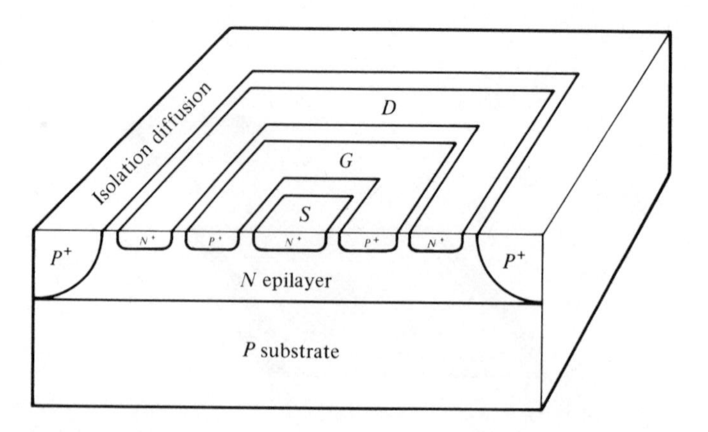

Fig. 1.3 Composite cross-sectional and top view of a modern J-FET.

the control or gating action performed by this terminal. The modern J-FET shown in Fig. 1.3, although somewhat different in physical appearance, is functionally equivalent to the original Shockley structure.

1.2 QUALITATIVE THEORY OF OPERATION

To establish the basic principles of J-FET operation, let us assume standard biasing conditions and consider the symmetrical, somewhat idealized, structure first proposed by Shockley. Given an n-type region between the source and drain, standard operational conditions prevail in the J-FET when the top and bottom gates are tied together, $V_G \leq 0$, and $V_D \geq 0$, as illustrated in Fig. 1.4. Note that with $V_G \leq 0$, the p-n junctions are always zero or reverse biased. Also, $V_D \geq 0$ ensures that the electrons in the n-region move from the source to the drain (in agreement with the naming of the S and D terminals). Our approach here will be to systematically change the terminal voltages and examine what is happening inside the device.

First suppose that the gate terminal is grounded, $V_G = 0$, and the drain voltage is increased in small steps starting from $V_D = 0$. At $V_D = 0$ (remember V_G is also zero) the device is in thermal equilibrium and about all one sees inside the structure are small depletion regions about the top and bottom p^+-n junctions (see Fig. 1.5(a)). The depletion regions extend, of course, primarily into the lightly doped, central n-region of the device. Stepping V_D to small positive voltages yields the situation pictured in Fig. 1.5(b). A current, I_D, begins to flow into the drain and through the nondepleted n-region sandwiched between the two p^+-n junctions. The nondepleted, current-carrying region, we might note, is referred to as the *channel*. For small V_D, the channel looks and acts like a simple resistor, and the resulting variation of I_D with V_D is linear (see Fig. 1.6(a)).

When V_D is increased above a few tenths of a volt, the device typically enters a new phase of operation. To gain insight into the revised situation, refer to Fig. 1.5(c), where an arbitrarily chosen potential of 5 V is assumed to exist at the drain terminal. Since the source is grounded, it naturally follows that somewhere in the channel the potential takes

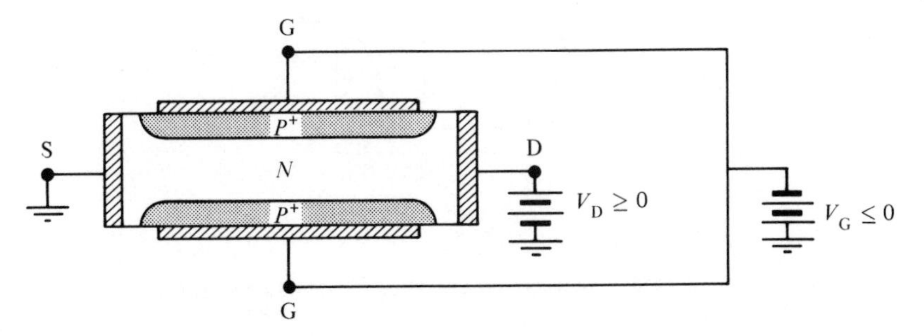

Fig. 1.4 Specification of the device structure and biasing conditions assumed in the qualitative analysis.

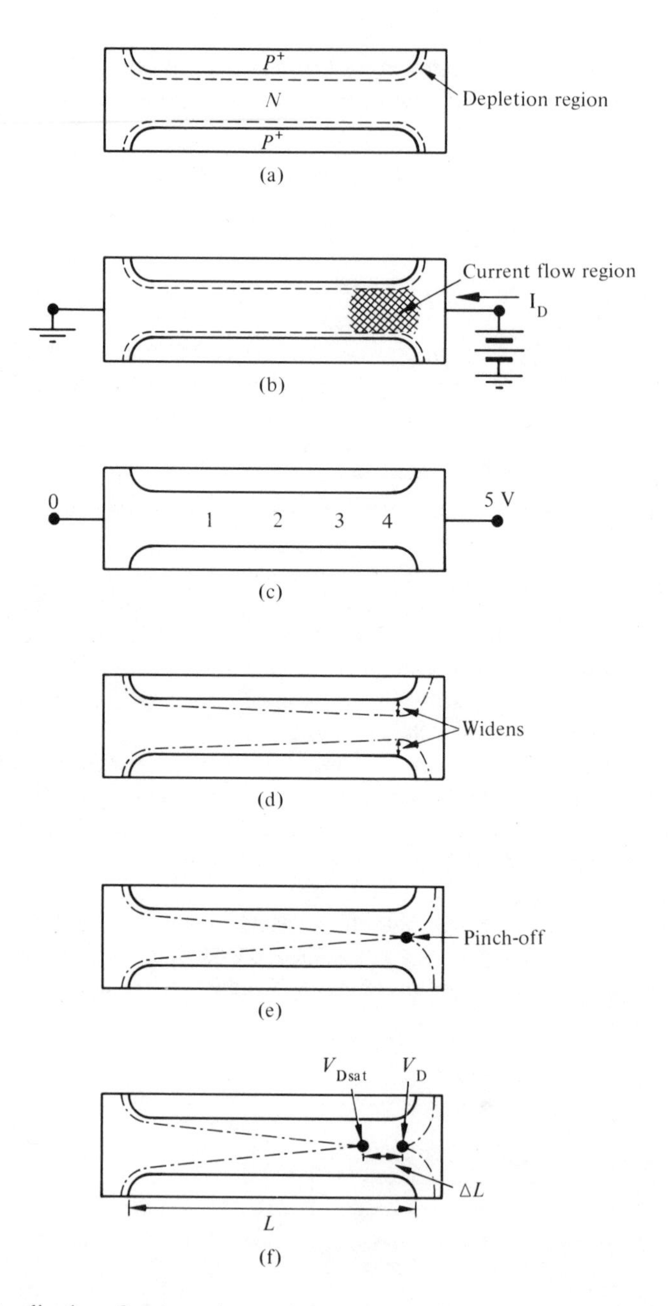

Fig. 1.5 Visualization of various phases of $V_G = 0$ J-FET operation. (a) Equilibrium ($V_D = 0$, $V_G = 0$); (b) small V_D biasing; (c) voltage drop down the channel for an arbitrarily assumed $V_D = 5V$; (d) channel narrowing under moderate V_D biasing; (e) pinch-off; (f) postpinch-off ($V_D > V_{Dsat}$).

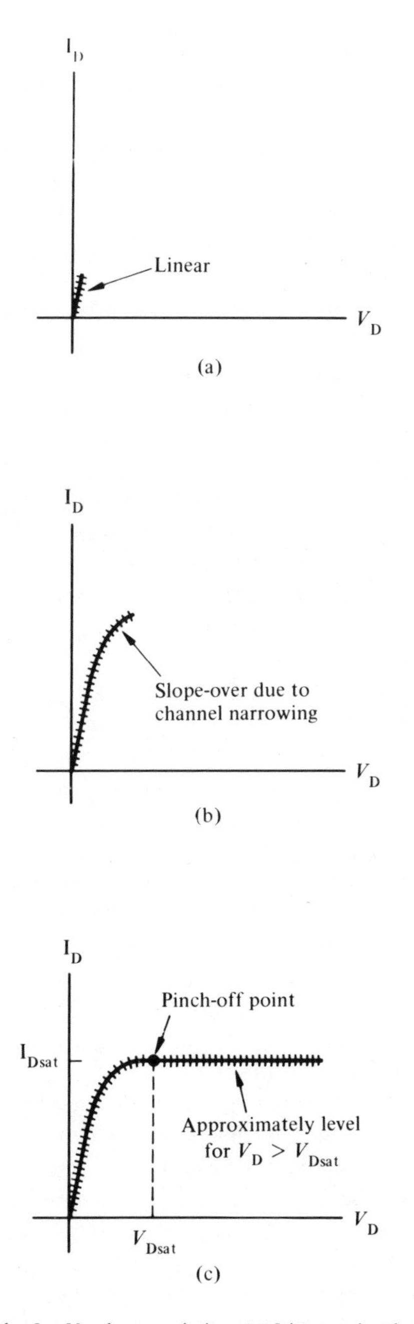

Fig. 1.6 General form of the I_D–V_D characteristics. (a) Linear, simple resistor, variation for very small drain voltages. (b) Slope-over at moderate drain biases due to channel narrowing. (c) Pinch-off and saturation for drain voltages in excess of V_{Dsat}.

on the values of 1, 2, 3, and 4 volts, with the potential increasing as one progresses from the source to the drain. The p^+ sides of the p^+-n junctions, however, are being held at zero bias. Consequently, the bias applied to the drain leads indirectly to a reverse biasing of the gate junctions and an increase in the junction depletion widths. Moreover, the top and bottom depletion regions progressively widen in going down the channel from the source to the drain (see Fig. 1.5d). Still thinking of the channel region (the nondepleted n-region) as a resistor, but no longer a simple resistor, one would expect the loss of conductive volume to increase the source-to-drain resistance and reduce the ΔI_D resulting from a given change in drain voltage. This is precisely the situation pictured in Fig. 1.6(b). The slope of the I_D–V_D characteristic decreases at larger drain biases because of the channel-narrowing effect.

Continuing to increase the drain voltage obviously causes the channel to narrow more and more, especially near the drain, until eventually the top and bottom depletion regions touch in the near vicinity of the drain, as pictured in Fig. 1.5(e). The complete depletion of the channel, touching of the top and bottom depletion regions, is an important special condition and is referred to as "*pinch-off*." When the channel pinches off inside the device, the slope of the I_D–V_D characteristic becomes approximately zero (see Fig. 1.6c), and the drain bias at the pinch-off point is given the special designation V_{Dsat}. For drain biases in excess of V_{Dsat}, the I_D–V_D characteristic saturates, that is, remains approximately constant at the I_{Dsat} value.

The statements presented without explanation in the preceding paragraph are totally factual. The I_D–V_D characteristic does level off or saturate when the channel pinches off. At first glance, however, the facts appear to run contrary to physical intuition. Should not pinch-off totally eliminate any current flow in the channel? How can one account for the fact that V_D voltages in excess of V_{Dsat} have essentially no effect on the drain current?

In answer to the first question, let us suppose I_D at pinch-off was identically zero. If I_D were zero, there would be no current in the channel at any point and the voltage down the channel would be the same as at $V_D = 0$, namely, zero everywhere. If the channel potential is zero everywhere, the p-n junctions would be zero biased and the channel in turn would be completely open from the source to the drain, clearly contradicting the initial assumption of a pinched-off channel. In other words, a current must flow in the J-FET to induce and maintain the pinched-off condition. Perhaps the conceptual difficulty often encountered with pinch-off arises from the need for a large current to flow through a depletion region. Remember, depletion regions are not totally devoid of carriers. Rather, the carrier numbers are just small compared to the background doping concentration (N_D or N_A) and may still approach densities $\sim 10^{12}/\text{cm}^3$ or greater. Moreover, the passage of large currents through a depletion region is not unusual in solid state devices. For example, a large current flows through the depletion region in a forward biased diode and through both depletion regions in a Bipolar Junction Transistor.

With regard to the saturation of I_D for drain biases in excess of V_{Dsat}, there is a very simple physical explanation. When the drain bias is increased above V_{Dsat} the pinched-off portion of the channel widens from just a point into a depleted channel section ΔL in extent. As shown in Fig. 1.5(f), the voltage on the drain side of the ΔL section is V_D, while the voltage on the source side of the section is V_{Dsat}. In other words, the applied drain voltage in excess of V_{Dsat}, $V_D - V_{Dsat}$, is dropped across the depleted section of the channel. Now, assuming $\Delta L \ll L$, the usual case, the source-to-pinch-off region of the device will be essentially identical in shape and will have the same endpoint voltages (zero and V_{Dsat}) as were present at the start of saturation. If the shape of a conducting region and the potential applied across the region do not change, then the current through the region must also remain invariant. This explains the approximate constancy of the drain current for postpinch-off biasing. [Naturally, if ΔL is comparable to L, then the same voltage drop (V_{Dsat}) will appear across a shorter channel section ($L - \Delta L$) and the postpinch-off I_D will increase perceptibly with increasing $V_D > V_{Dsat}$. This effect is especially noticeable in short channel (small L) devices.]

Another approach to explaining the saturation of the I_D–V_D characteristics makes use of an analogous situation in everyday life; namely, a waterfall. As everyone knows, the water flow rate over a falls is controlled not by the height of the falls, but by the flow rate down the rapids leading to the falls. Thus, assuming an identical rapids region, the water flow rate at the bottom of the two falls pictured in Fig. 1.7 is precisely the same, even though the heights of the falls are different. The rapids region is of course analogous to the source side of the channel in the J-FET, the falls proper corresponds to the pinched-off ΔL section at the drain end of the channel, and the height of the falls corresponds to the $V_D - V_{Dsat}$ potential drop across the ΔL section.

Thus far we have established the expected variation of I_D with V_D when $V_G = 0$. To complete the discussion we need to investigate the operation of the J-FET when $V_G < 0$. $V_G < 0$ operation, it turns out, is very similar to $V_G = 0$ operation with three minor modifications. First, if $V_G < 0$ the top and bottom p^+-n junctions are reverse biased even when $V_D = 0$. A reverse bias on the junctions increases the width of the depletion regions and shrinks the $V_D = 0$ lateral width of the channel. Consequently, the resistance of the channel increases at a given V_D value and the linear portion of the I_D–V_D characteristic exhibits a smaller slope when $V_G < 0$ (see Fig. 1.8(a)). Secondly, because the channel is narrower at $V_D = 0$, the channel also becomes pinched-off at a smaller drain bias. Therefore, as pictured in Fig. 1.8(b), V_{Dsat} and I_{Dsat} when $V_G < 0$ are smaller than V_{Dsat} and I_{Dsat} when $V_G = 0$. Finally, note that for sufficiently negative V_G biases it is possible

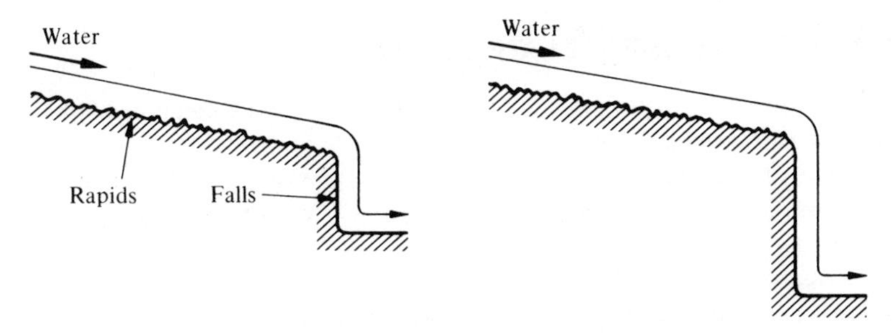

Fig. 1.7 The waterfalls analogy.

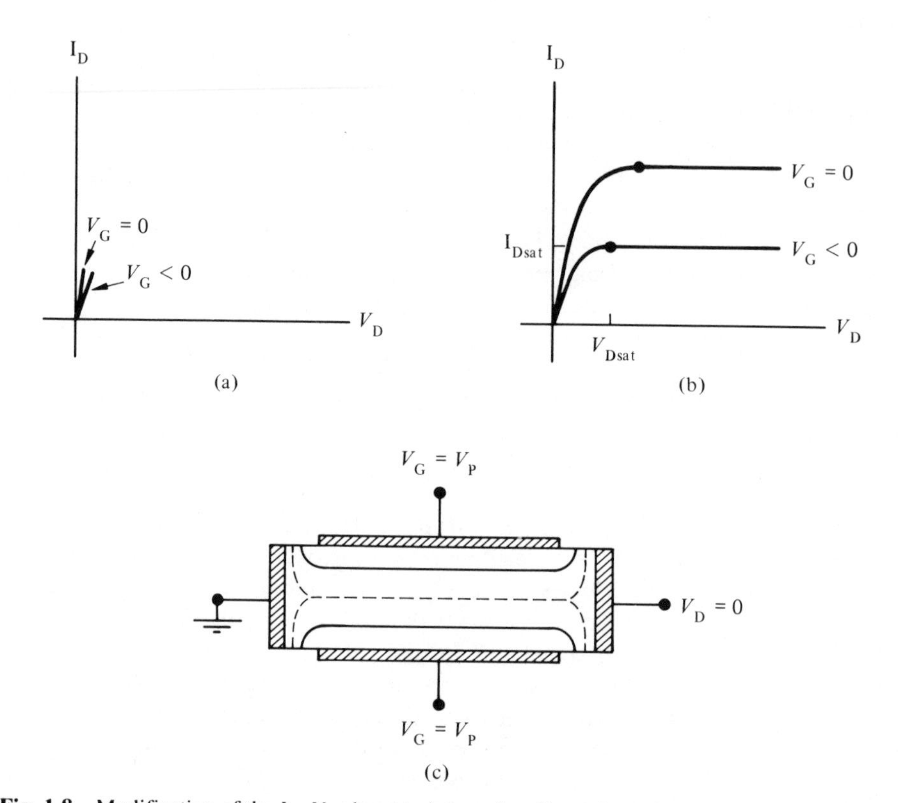

Fig. 1.8 Modification of the I_D–V_D characteristics when $V_G < 0$. (a) Decrease in the linear slope of the characteristics for small drain voltages. (b) Decrease in the saturation current and saturation drain voltage. (c) Gate pinch-off.

to deplete the entire channel even with $V_D = 0$ (see Fig. 1.8c). The gate bias, $V_G = V_P$, where the gate voltage first totally depletes the entire channel with V_D set equal to zero, is referred to as the pinch-off gate voltage. For $V_G \leq V_P$ the drain current is identically zero for all drain biases.*

1.3 QUANTITATIVE I_D–V_D RELATIONSHIPS

Wanted: a quantitative expression for the drain current as a function of the terminal voltages; that is, $I_D = I_D(V_D, V_G)$.

Device Specification. The precise device structure, dimensions, and assumed coordinate orientations are as specified in Fig. 1.9. The y-axis is directed down the channel from

---

*If the drain bias is made very large, the p^+-n junctions in the vicinity of the drain will eventually break down, leading to a very rapid increase in I_D with V_D for any gate bias. This breakdown region has been omitted from all of the theoretical characteristics sketched herein.

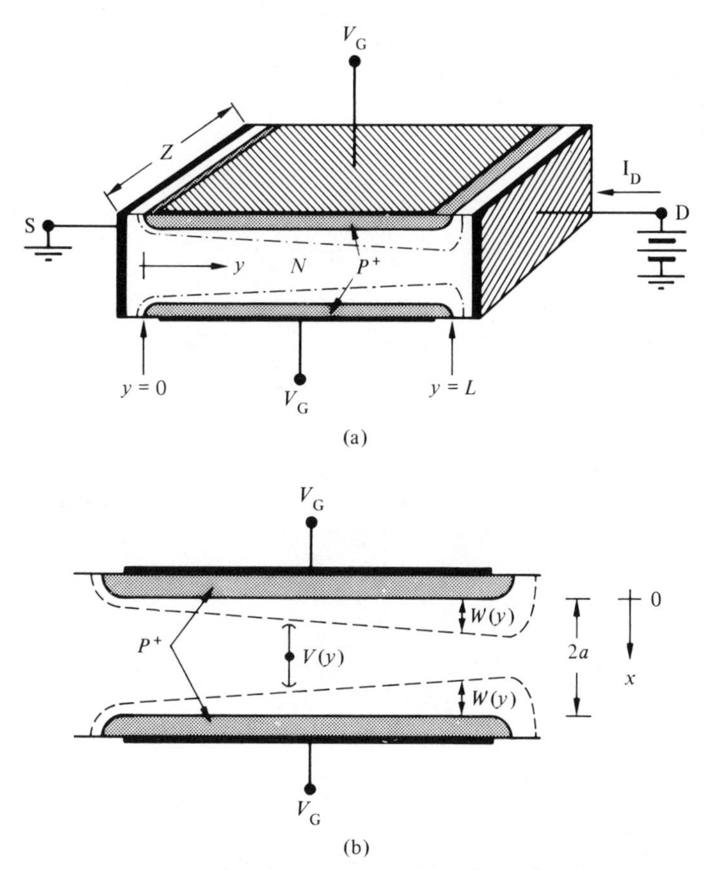

Fig. 1.9 Device structure, dimensions, and coordinate orientations assumed in the quantitative analysis. (a) Overall diagram. (b) Expanded view of the channel region.

the source to the drain while the x-coordinate is oriented normal to the p^+-n metallurgical junctions. L is the channel length, Z is the p^+-n junction width, and $2a$ is the distance between the top and bottom metallurgical junctions. Note that $y = 0$ and $y = L$ are slightly removed from the source and drain contacts, respectively, $V(y)$ is the potential and $W(y) = W_{top}(y) = W_{bottom}(y)$ is the junction depletion width at an arbitrary point y in the channel. $W(y)$ lies, of course, almost totally in the n-region because of the p^+-n nature of the junctions.

Basic Assumptions. (1) The junctions are p^+-n *step* junctions and the n-region is uniformly doped with a donor concentration equal to N_D. (2) The device is structurally symmetrical about the $x = a$ plane as shown in Fig. 1.9, and the symmetry is maintained by operating the device with the same V_G bias applied to the top and bottom gates. (3) Current flow is confined to the nondepleted portions of the n-region and is directed exclusively in the

y-direction. (4) $W(y)$ can be increased to at least a width a without inducing breakdown in the p^+-n junctions. (We implicitly assumed this to be the case in the qualitative discussion.) (5) Voltage drops from the source to $y = 0$ and from $y = L$ to the drain are negligible.

For drain and gate voltages below pinch-off, $0 \leq V_D \leq V_{Dsat}$ and $0 \geq V_G \geq V_P$, the derivation of the desired I_D–V_D relationship proceeds as follows: In general one can write

$$\mathbf{J}_N = q\mu_n n\mathscr{E} + qD_N\nabla n \tag{1.1}$$

Within the conducting channel $n \simeq N_D$ and the current is flowing almost exclusively in the y-direction. Moreover, with $n \simeq N_D$, the diffusion component of the current $(qD_N\nabla n)$ should be relatively small. Under the cited conditions Eq. (1.1) reduces to

$$J_N = J_{Ny} = q\mu_n N_D \mathscr{E}_y = -q\mu_n N_D \frac{dV}{dy} \qquad \text{(in the conducting channel)} \tag{1.2}$$

Since there are no carrier sinks or sources in the device, the current flowing through any cross-sectional plane within the channel must be equal to I_D. Thus, integrating the current density over the cross-sectional area of the conducting channel at an arbitrary point y yields

$$I_D = -\iint J_{Ny}\, dx\, dz = -Z\int_{W(y)}^{2a-W(y)} J_{Ny}\, dx = 2Z\int_{W(y)}^{a} q\mu_n N_D \frac{dV}{dy} dx \tag{1.3a}$$

$$= 2qZ\mu_n N_D a \frac{dV}{dy}\left(1 - \frac{W}{a}\right) \tag{1.3b}$$

The minus sign appears in the general formula for I_D because I_D is defined to be positive in the $-y$ direction. Use was also made of the fact that the structure is symmetrical about the $x = a$ plane.

Remembering that I_D is independent of y, one can recast Eq. (1.3b) into a more useful form by integrating I_D over the length of the channel. Specifically,

$$\int_0^L I_D\, dy = I_D L = 2qZ\mu_n N_D a \int_{V(0)=0}^{V(L)=V_D} \left[1 - \frac{W(V)}{a}\right] dV \tag{1.4}$$

or

$$I_D = \frac{2qZ\mu_n N_D a}{L} \int_0^{V_D} \left[1 - \frac{W(V)}{a}\right] dV \tag{1.5}$$

To proceed any further we need an analytical expression for W as a function of V. It should be recognized that the electrostatic problem inside the J-FET is really two-dimensional in nature. To obtain W as a function of V exactly would necessitate the solution of Poisson's equation taking into account both the x and y variation of the electrostatic variables. The apparent impasse here is circumvented, and an approximate expression for W as a function of V obtained, by invoking the *gradual channel approximation*. This approximation, encountered frequently in the analysis of field-effect devices, is based on the assumption that the rate of change of the electrostatic variables (potential, electric field, etc.) in, say, the y-direction is relatively slow compared to the rate of change of the same variables in the x-direction. (Such is clearly the case in the J-FET, for example, if $L \gg a$.) Under the cited assumption, the y-direction dependence is then totally neglected and the electrostatic variables at each point y computed using the simple one-dimensional analysis.

For the specific problem at hand, invoking the gradual channel approximation simply means that we employ the standard one-dimensional p-n junction depletion width expression derived in an earlier volume. Thus, for the assumed p^+-n step junctions,

$$W(V) = \left[\frac{2\varepsilon}{qN_D}(V_{bi} - V_A)\right]^{1/2} = \left[\frac{2\varepsilon}{qN_D}(V_{bi} + V - V_G)\right]^{1/2} \tag{1.6}$$

where, as is evident from Fig. 1.9(b), $V_A = V_G - V(y)$ is the applied potential drop across the junction at a given point y. It is next convenient to note that $W \to a$ when $V_D = 0$ ($V = 0$) and $V_G = V_P$. Thus substituting into Eq. (1.6) yields

$$a = \left[\frac{2\varepsilon}{qN_D}(V_{bi} - V_P)\right]^{1/2} \tag{1.7}$$

and

$$\frac{W(V)}{a} = \left(\frac{V_{bi} + V - V_G}{V_{bi} - V_P}\right)^{1/2} \tag{1.8}$$

Finally, substituting $W(V)/a$ from Eq. (1.8) into Eq. (1.5) and performing the indicated integration, one obtains

$$\boxed{I_D = \frac{2qZ\mu_n N_D a}{L}\left\{V_D - \frac{2}{3}(V_{bi} - V_P)\left[\left(\frac{V_D + V_{bi} - V_G}{V_{bi} - V_P}\right)^{3/2} - \left(\frac{V_{bi} - V_G}{V_{bi} - V_P}\right)^{3/2}\right]\right\} \quad \text{for } 0 \le V_D \le V_{Dsat}; \quad V_P \le V_G \le 0} \tag{1.9}$$

It should be reemphasized that the foregoing development and Eq. (1.9), in particular, apply only below pinch-off. In fact, the computed I_D versus V_D for a given V_G actually begins to decrease if V_D values in excess of V_{Dsat} are inadvertently substituted into Eq. (1.9). As pointed out in the qualitative discussion, however, I_D is approximately constant if V_D exceeds V_{Dsat}. To first order, then, the post-pinch-off portion of the characteristics can be modeled by simply setting

$$I_{D|V_D > V_{Dsat}} = I_{D|V_D = V_{Dsat}} \equiv I_{Dsat} \tag{1.10a}$$

or

$$I_{Dsat} = \frac{2qZ\mu_n N_D a}{L}\left\{V_{Dsat} - \frac{2}{3}(V_{bi} - V_P)\left[\left(\frac{V_{Dsat} + V_{bi} - V_G}{V_{bi} - V_P}\right)^{3/2} - \left(\frac{V_{bi} - V_G}{V_{bi} - V_P}\right)^{3/2}\right]\right\} \tag{1.10b}$$

The I_{Dsat} relationship can be simplified somewhat by noting that pinch-off at the drain end of the channel implies $W \rightarrow a$ when $V(L) = V_{Dsat}$. Therefore, from Eq. (1.6),

$$a = \left[\frac{2\varepsilon}{qN_D}(V_{bi} + V_{Dsat} - V_G)\right]^{1/2} \tag{1.11}$$

Comparing Eqs. (1.7) and (1.11) one concludes

$$\boxed{V_{Dsat} = V_G - V_P} \tag{1.12}$$

and

$$\boxed{I_{Dsat} = \frac{2qZ\mu_n N_D a}{L}\left\{V_G - V_P - \frac{2}{3}(V_{bi} - V_P)\left[1 - \left(\frac{V_{bi} - V_G}{V_{bi} - V_P}\right)^{3/2}\right]\right\}} \tag{1.13}$$

Theoretical $I_D - V_D$ characteristics computed using Eqs. (1.9) and (1.13) are presented in Fig. 1.10. For comparison purposes a sample set of experimental characteristics are displayed in Fig. 1.11. Generally speaking, the theory does a reasonably adequate job of modeling the experimental observations. A somewhat improved agreement between experiment and theory can be achieved by lifting the assumption of negligible voltage drops in the regions of the device between the active channel and the source/drain contacts

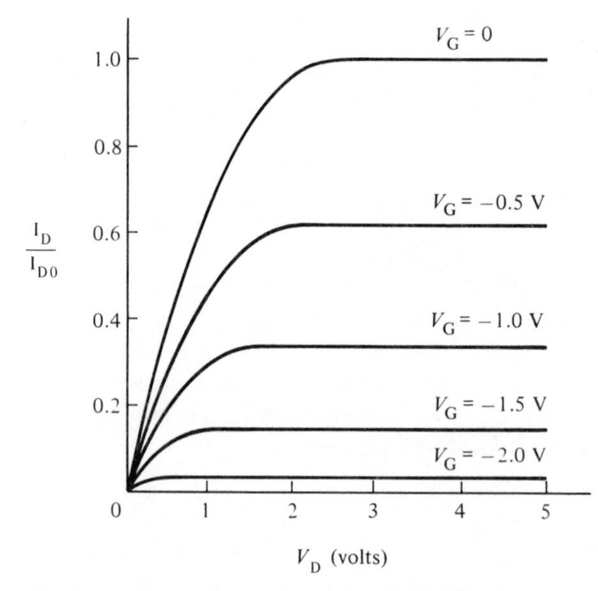

Fig. 1.10 Normalized theoretical I_D–V_D characteristics assuming $V_{bi} = 1$ V and $V_P = -2.5$ V. $I_{DO} = I_{Dsat|V_G=0}$.

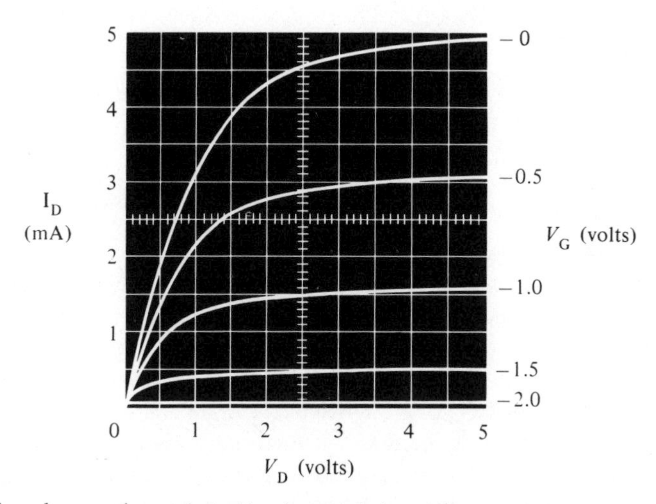

Fig. 1.11 Sample experimental I_D–V_D characteristics. (Characteristics were derived from a TI 2N3823 *n*-channel J-FET.)

(see Fig. 1.12). We leave it to the reader as an exercise to establish the resulting theoretical modifications. Finally, it should be pointed out that in saturation most J-FET characteristics can be closely modeled by the simple relationship

$$\boxed{I_{Dsat} = I_{D0}(1 - V_G/V_P)^2} \qquad \text{where } I_{DO} = I_{Dsat|V_G=0} \tag{1.14}$$

Although appearing totally different when compared to Eq. (1.13), the semiempirical "square-law" relationship of Eq. (1.14) yields similar numerical results and is much simpler to employ in circuit calculations where the J-FET is viewed as a "black box." Equation (1.13) (or similar first-principle result), on the other hand, is indispensable if one wishes to investigate the dependence of the J-FET characteristics on temperature, channel doping, or some other fundamental device parameter.

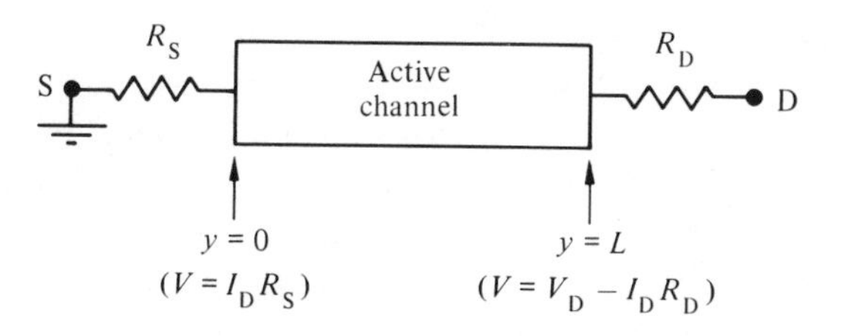

Fig. 1.12 Modified model for the J-FET including the resistance of the semiconductor regions between the ends of the active channel and the source/drain terminals.

1.4 ac RESPONSE

The ac response of the J-FET, routinely expressed in terms of the J-FET's small signal equivalent circuit, is most conveniently established by considering the two-port network shown in Fig. 1.13. Herein we will restrict our considerations to low operational frequencies where capacitive effects may be neglected.

Let us begin by examining the device input. Under standard dc biasing conditions, the input port between the gate and source is connected across a reverse biased diode on the inside of the structure. A reverse biased diode, however, behaves (to first order) like an open circuit at low frequencies. It is standard practice, therefore, to model the input to the J-FET by an open circuit.

At the output port the dc drain current has already been established to be a function of V_D and V_G; that is, $I_D = I_D(V_D, V_G)$. When the ac drain and gate potentials, v_d and v_g, are respectively added to the dc drain and gate terminal voltages, V_D and V_G, the drain current through the structure is of course modified to $I_D(V_D, V_G) + i_d$, where i_d is the ac component of the drain current. Provided the device can follow the ac changes in potential quasi-statically,* which is assumed to be the case at low operational frequencies, one can state

$$i_d + I_D(V_D, V_G) = I_D(V_D + v_d, V_G + v_g) \tag{1.15a}$$

and

$$i_d = I_D(V_D + v_d, V_G + v_g) - I_D(V_D, V_G) \tag{1.15b}$$

Expanding the first term on the right-hand side of Eq. (1.15b) in a Taylor series and keeping only through first-order terms in the expansion (higher-order terms are negligible), one obtains

$$i_d = \left.\frac{\partial I_D}{\partial V_D}\right|_{V_G} v_d + \left.\frac{\partial I_D}{\partial V_G}\right|_{V_D} v_g \tag{1.16a}$$

or

$$i_d = g_d v_d + g_m v_g \tag{1.16b}$$

where

$$g_d \equiv \left.\frac{\partial I_D}{\partial V_D}\right|_{V_G = \text{constant}} \quad \text{the drain or channel conductance} \tag{1.17a}$$

$$g_m \equiv \left.\frac{\partial I_D}{\partial V_G}\right|_{V_D = \text{constant}} \quad \text{transconductance or mutual conductance} \tag{1.17b}$$

---

*The term "quasi-static" is used to describe situations where the time varying state of a system at any given instant is essentially indistinguishable from the dc state that would be achieved under equivalent biasing conditions.

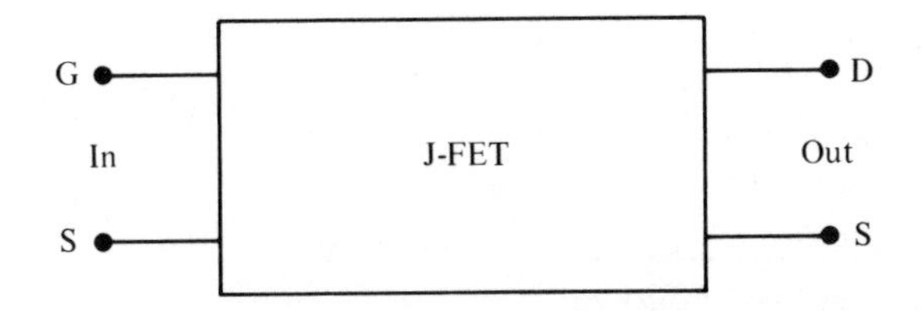

Fig. 1.13 The J-FET viewed as a two-port network.

Equation (1.16b) may be viewed as the ac-current node equation for the drain terminal and, by inspection, leads to the output portion of the circuit displayed in Fig. 1.14. Since, as concluded earlier, the gate-to-source or input portion of the device is simply an open circuit, Fig. 1.14 then is the desired small signal equivalent circuit characterizing the low frequency ac response of the J-FET. Note that for field-effect transistors the g_m parameter plays a role analogous to the α's and β's in the modeling of bipolar junction transistors. g_d, as its name indicates, may be viewed as either the device output admittance or the ac conductance of the channel between the source and drain. Explicit g_d and g_m relationships obtained by direct differentiation of Eqs. (1.9) and (1.13), using the Eq. (1.17) definitions, are catalogued in Table 1.1. The $g_d = 0$ result in Table 1.1 for operation of the device under saturation conditions is of course consistent with the theoretically zero slope of the $I_D - V_D$ characteristics when $V_D \geq V_{Dsat}$.

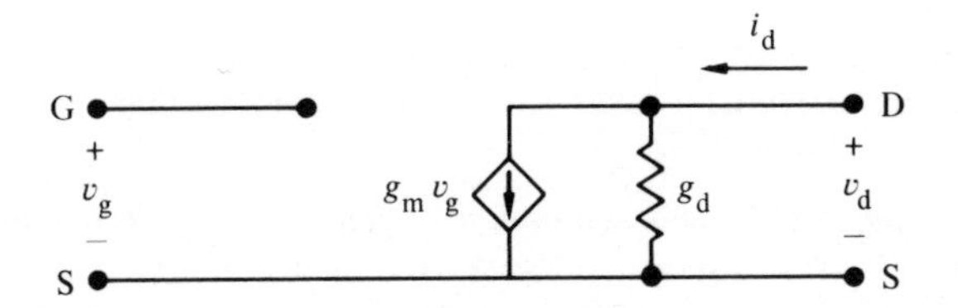

Fig. 1.14 Small signal equivalent circuit characterizing the low-frequency ac response of the J-FET.

Table 1.1 J-FET Small Signal Parameters. Entries in the table were obtained by direct differentiation of Eqs. (1.9) and (1.13). $G_0 \equiv 2qZ\mu_n N_D a/L$.

Below Pinch-off ($V_D \leq V_{Dsat}$)	Post-Pinch-off ($V_D \geq V_{Dsat}$)
$g_d = G_0\left[1 - \left(\frac{V_D + V_{bi} - V_G}{V_{bi} - V_P}\right)^{1/2}\right]$	$g_d = 0$
$g_m = G_0\left[\left(\frac{V_D + V_{bi} - V_G}{V_{bi} - V_P}\right)^{1/2} - \left(\frac{V_{bi} - V_G}{V_{bi} - V_P}\right)^{1/2}\right]$	$g_m = G_0\left[1 - \left(\frac{V_{bi} - V_G}{V_{bi} - V_P}\right)^{1/2}\right]$

PROBLEMS

1.1 Answer the following questions as concisely as possible.

(a) Define "field-effect."

(b) Precisely what is the "channel" in J-FET terminology?

(c) What is meant by the term "pinch-off"?

(d) For a p-channel device (a device with n^+-p gating junctions and a p-region between the source and drain), does the drain current flow into or out of the drain contact under normal operational conditions? Explain.

(e) What is the "gradual channel approximation"?

(f) What is the mathematical definition of the drain conductance? of the transconductance?

(g) Draw the small signal equivalent circuit characterizing the low-frequency ac response of a *long*-channel J-FET under *saturation* conditions. (Assume the dc characteristics of the device are similar to those shown in Fig. 1.10.)

1.2 As shown in Fig. 1.5(c), the variation in voltage down the length of the J-FET channel is typically a nonlinear function of position when V_D is greater than a few tenths of a volt.

(a) Derive an expression which can be used to compute the point in the channel (that is, y/L) where a given channel voltage $0 \le V(y) \le V_D$ is expected to occur. *Hint*: Let $L \to y$ and $V_D \to V(y)$ in Eq. (1.4), solve for y, evaluate the remaining integral, and then form the ratio y/L from your result.

(b) Assuming $V_G = 0$, $V_D = 5$ V, $V_{bi} = 1$ V, and $V_P = -8$ V, calculate y/L for $V(y) = 1, 2, 3,$ and 4 V. How do the computed y/L values compare with the positioning of the voltages shown in Fig. 1.5(c)?

1.3 Noting that the Eq. (1.9) expression for I_D maximizes at $V_D = V_{Dsat}$ for a given V_G, apply the standard mathematical procedure for determining extrema points of a function to derive Eq. (1.12) directly from Eq. (1.9).

1.4 Compare the I_{Dsat}/I_{D0} values obtained from the square-law relationship [Eq. (1.14)] with the Eq. (1.13) based I_{Dsat}/I_{D0} values shown in Fig. 1.10. Comment on the comparison.

1.5 In this problem we wish to explore the effect of non-negligible voltage drops in the regions of the J-FET between the active channel and the source/drain contacts (see Fig. 1.12).

(a) Appropriately revise Eqs. (1.4) and (1.5) setting $V(0) = I_D R_S$ and $V(L) = V_D - I_D R_D$.

(b) Evaluate the revised Eq. (1.5) integral to obtain an expression analogous to Eq. (1.9). (Make no attempt to solve the resulting equation for I_D.)

(c) Establish expressions analogous to Eqs. (1.12) and (1.13).

(d) Verify that the results established in parts (b) and (c) could have been obtained by simply replacing V_D with $V_D - I_D(R_S + R_D)$ and V_G with $V_G - I_D R_S$ in the corresponding text expressions.

1.6 Suppose, as shown in Fig. P1.6, the bottom gate lead of a J-FET is tied to the source and grounded.

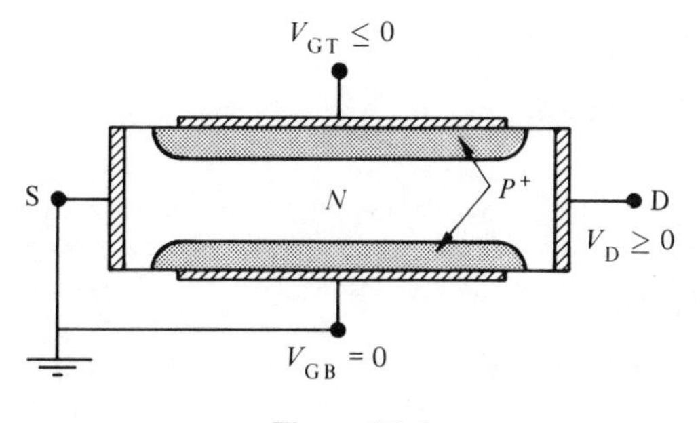

Figure P1.6

(a) Sketch an outline of the depletion regions inside the $V_{GB} = 0$ device when V_{GT} is made sufficiently large to pinch-off the channel with $V_D = 0$.

(b) If V_P (the pinch-off gate voltage) $= -8$ V when the two gates are tied together, $V_{bi} = 1$ V, and assuming p^+-n step junctions, determine V_{PT} (the top gate pinch-off voltage) when $V_{GB} = 0$ and $V_D = 0$. Is your answer here consistent with the sketch in part (a)? Explain.

(c) Assuming $V_{PT} < V_{GT} < 0$, sketch an outline of the depletion regions inside the device when the *drain* voltage is increased to the pinch-off point.

(d) Derive an expression which specifies V_{Dsat} in terms of V_{PT}, V_{bi}, and V_{GT} for $V_{GB} = 0$ operation. (Your answer should contain only voltages. Make no attempt to actually solve for V_{Dsat}.)

(e) In light of your answers to parts (c) and (d), will V_{Dsat} for $V_{GB} = 0$ operation be greater than or less than the V_{Dsat} for $V_{GB} = V_{GT}$ operation? Explain.

(f) Derive an expression for I_D as a function of V_D and V_{GT} analogous to Eq. (1.9).

1.7 A J-FET is constructed with the *gate-to-gate* doping profile shown in Fig. P1.7. Specifically assume the p^+-region doping is much greater than the maximum n-region doping and make other obvious assumptions as required.

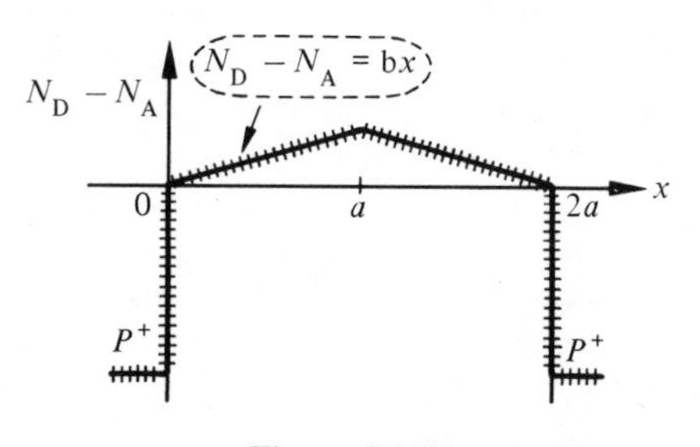

Figure P1.7

(a) Starting with Poisson's equation, derive an expression for the *left-hand* junction depletion width (W). Let V_A be the applied voltage drop across the junction. (If necessary refer back to Section 2.3 in Vol. II.)

(b) Neglecting μ_n's doping dependence and assuming the left-hand and right-hand gates are tied together, appropriately modify the text J-FET analysis to obtain the below-pinch-off I_D–V_D relationship for this linearly graded junction. (*Caution*: more than the $W(V)/a$ expression must be modified.)

1.8 In this problem you will be asked various questions concerning the device shown in Fig. P1.8. The device, which might be called a compositristor (composite transistor) is formed from a uniformly doped n-type bar. Ohmic contacts are made to the top and bottom of the bar and are connected to the outside world through leads D and B, respectively. p^+-n abrupt junctions are formed on the two sides of the bar and are connected to the outside world through contacts E and C. As shown in the figure, d is the separation between the two p^+ regions and L is the lateral length of the p^+ regions. *To receive full credit you must indicate your reasoning* in addition to answering each of the following.

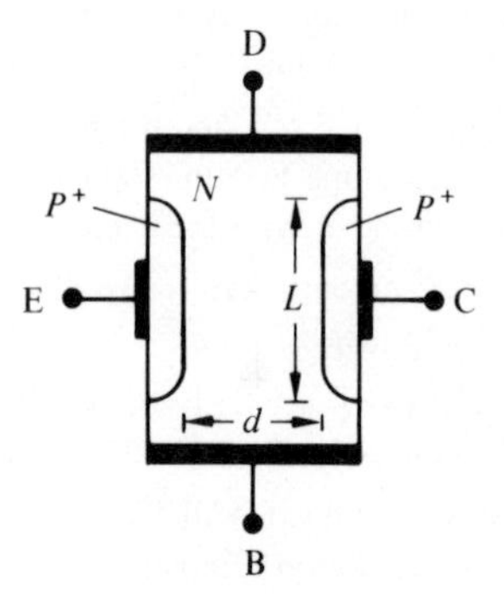

Figure P1.8

(a) *Given*: D–B tied together; $d \ll L_P$, where L_P is the minority carrier diffusion length in the n-region; $V_{EB} > 0$; $V_{CB} < 0$.
Problem: Sketch the current (I_C) flowing *out* of the C contact as a function of V_{CB} if I_E is held constant at various different values.

(b) *Given*: E–C tied together; D–B tied together; $d \gg L_P$.
Problem: Sketch the current flowing into the E–C leads as a function of the voltage applied from E–C to D–B.

(c) *Given*: E–C tied together; $d < 2W_{max}$, where W_{max} is the maximum depletion width before avalanching takes place for each of the p^+-n junctions; L is almost equal to the total length of the bar between D and B; $V_{DB} > 0$; $V_{EB} < 0$.
Problem: Sketch the current (I_D) flowing into the contact D as a function of the V_{DB} voltage if V_{EB} is held constant at various different values.

2 / Ideal MOS Device Statics

The metal–oxide (SiO_2)–semiconductor (Si), or MOS structure, is without a doubt the heart of present-day electronics. Even ostensibly *pn*-junction type devices incorporate the MOS structure in some functional and/or physical manner. A quasi-MOS device, as noted in Chapter 1, was first proposed in the 1920s. The dawn of modern history, however, is generally attributed to D. Kahng and M. M. Atalla who filed patents on the Si–SiO_2 based field-effect transistor in 1960. The MOS designation, it should be noted, is specifically reserved for the technological giant, the metal–SiO_2–Si system. The more general designation, Metal–Insulator–Semiconductor (MIS), is used to identify similar structures composed of an insulator other than SiO_2 or a semiconductor other than Si.

In this chapter we initiate the presentation of MOS device "basics." Of primary concern is the internal status of the structure—particularly the charge, electric field, and band bending inside the semiconductor—under static conditions and assuming the structure is ideal. After a precise specification of the "ideal" structure, energy band diagrams and charge diagrams are utilized to qualitatively visualize the static state. Band bending inside the semiconductor is next quantitatively analyzed and subsequently related to the voltage applied to the metallic gate. As a whole this chapter also serves as an introduction to the highly specialized and rather extensive MOS terminology.

2.1 THE IDEAL MOS STRUCTURE

The simplest of all MOS devices, the MOS-Capacitor, is pictured in Fig. 2.1. The MOS-C, a fundamental building block in more complex MOS device structures, is a two terminal device composed of a thin ($\sim 0.1\mu = 10^{-5}$cm thick) SiO_2 layer sandwiched between a silicon substrate and a metallic field plate. The most common field plate materials are aluminum and heavily doped polycrystalline silicon.* A second metallic layer present

*Heavily doped Si is metallic in nature. Poly-Si gates, used extensively in complex MOS device structures, are deposited by a chemical-vapor process and are then heavily doped by either phosphorus or boron diffusion.

along the back or bottom side of the semiconductor provides a contact between the silicon and the outside world. The terminal connected to the field plate and the field plate itself are referred to as the gate; the silicon-side terminal, which is normally grounded, is simply called the back contact.

The ideal MOS structure assumed throughout this chapter has the following explicit properties: (1) the metallic gate is sufficiently thick so that it can be considered an equipotential region under ac as well as dc biasing conditions; (2) the oxide is a *perfect insulator* with *zero current* flowing through the oxide layer under *all* static biasing conditions; (3) there are no charge centers located in the oxide or at the oxide-semiconductor interface; (4) the semiconductor is uniformly doped; (5) the semiconductor is sufficiently thick so that, regardless of the applied gate potential, a field-free region (the so-called Si "bulk") is encountered before reaching the back contact; (6) an *ohmic* contact has been established between the semiconductor and the metal on the back side of the device; (7) the MOS-C is a one-dimensional structure with all variables taken to be a function only of the x-coordinate (see Fig. 2.1); (8) it is assumed that $\Phi_M = \chi + (E_c - E_F)_\infty$, where Φ_M, χ, and $(E_c - E_F)_\infty$ are material parameters (energies) defined in Fig. 2.2. (A discussion of these material parameters is presented in the next section.)

All of the listed idealizations can be approached in practice and the ideal MOS structure is a fairly realistic animal. For example, the resistivity of SiO_2 can be as high as 10^{18} ohm-cm, and the dc leakage current through the layer is indeed negligible for typical oxide thicknesses and applied voltages. Moreover, even very thin gates can be considered equipotential regions and ohmic back contacts are quite easy to achieve in practice. Similar statements can be made concerning most of the other idealizations. Special note, however, should be made of idealization #8. The $\Phi_M = \chi + (E_c - E_F)_\infty$ requirement could be omitted and will in fact be eliminated in a later chapter. The requirement has been included here only to avoid unnecessary complications in the initial description of the static state.

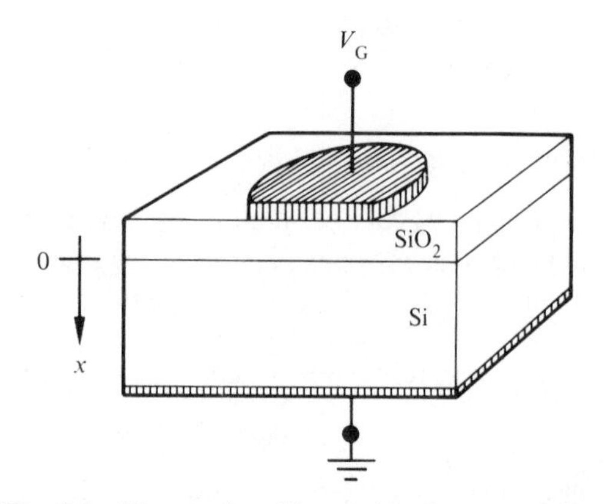

Fig. 2.1 The metal–oxide–semiconductor capacitor.

2.2 VISUALIZATION OF THE STATIC STATE

2.2.1 The Energy Band Diagram

The energy band diagram is an indispensable aid in visualizing the internal status of the MOS structure under static biasing conditions. Our first task is to construct the diagram appropriate for the ideal MOS structure under zero bias conditions and to establish general ground rules to be applied in modifying the diagram as a function of bias.

Figure 2.2 shows the surface-included energy band diagrams for the individual components of the MOS structure. In each case the abrupt termination of the diagram in a vertical line is meant to imply a surface. The ledge at the top of the vertical line, known as the vacuum level, denotes the minimum energy an electron must possess to completely free itself from the material. The energy difference between the vacuum level and the Fermi energy in a metal is known as the metal workfunction, Φ_M. In the semiconductor the height of the surface energy barrier is specified in terms of the electron affinity, χ, the energy difference between the vacuum level and the conduction band edge at the surface. χ is used instead of $E_{vacuum} - E_F$ because the latter quantity is not a constant in semiconductors, but varies as a function of doping and band bending near the surface. The remaining component, the insulator, is in essence modeled as an intrinsic wide-gap semiconductor where the surface barrier is again specified in terms of the electron affinity.

Formation of the unified equilibrium diagram is achieved by conceptually bringing the metal and semiconductor together until they are a distance x_o apart and then inserting the insulator of thickness x_o into the empty space between the metal and semiconductor components. Since the Fermi level must line up in any structure under equilibrium conditions (see Section 3.2, Volume I), and because we have specified $\Phi_M = \chi + (E_c - E_F)_\infty$, the vacuum levels of the M and S components are always in

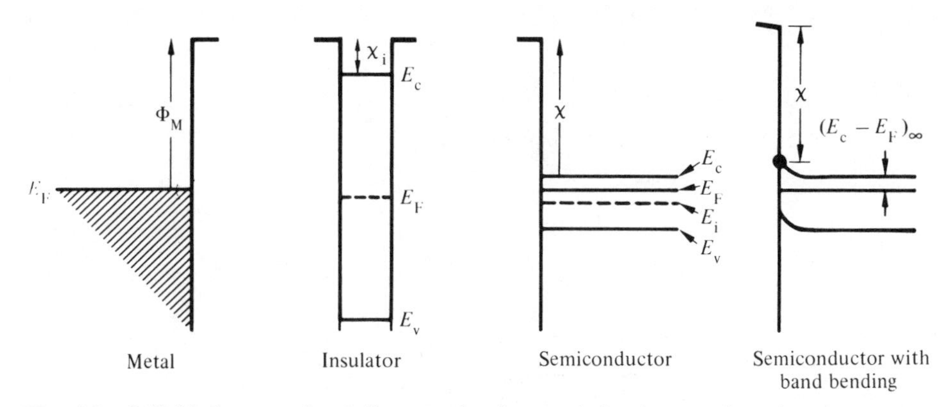

Fig. 2.2 Individual energy band diagrams for the metal, insulator, and semiconductor components of the MOS structure. The diagram labeled "semiconductor with band bending" defines $(E_c - E_F)_\infty$ and shows χ to be invariant with band bending. The value, χ, it should be emphasized, is measured relative to E_c at the semiconductor surface.

perfect alignment and no fields are induced in the system as the conducting materials are brought close together. Moreover, with a zero field in the x_o gap, inserting the insulator simply lowers the barrier slightly between the metal and semiconductor components. Thus the equilibrium energy band diagram for the ideal MOS structure is concluded to be of the form pictured in Fig. 2.3.

Seeking next to establish ground rules in working with the MOS energy band diagram, let us consider what happens in general to the diagram when the back side of the device is grounded and a nonzero static bias, V'_G, is applied to the gate. The prime in V'_G, it should be interjected, indicates at a glance that one is referring to the ideal structure; the unprimed symbol V_G is specifically reserved for the gate voltage applied to an actual MOS-C. With V'_G applied we note first of all that the *semiconductor Fermi energy is unaffected by the bias and remains invariant (level on the diagram) as a function of position*. This is a direct consequence of the assumed zero current flow through the structure under all static biasing conditions. In essence, the semiconductor always remains in equilibrium independent of the bias applied to the MOS-C gate. Secondly, as in a *pn* junction, the applied bias separates the Fermi energies at the two ends of the structure by an amount equal to qV'_G; that is,

$$E_F\,(\text{metal}) - E_F\,(\text{semiconductor}) = -qV'_G \tag{2.1}$$

Conceptually, the metal and semiconductor Fermi levels may be thought of as "handles" connected to the outside world. In applying a bias, one grabs onto the handles and rearranges the relative up-and-down positioning of the Fermi levels. The back contact is grounded and the semiconductor-side handle therefore remains fixed in position. The metal-side handle, on the other hand, is moved downward if $V'_G > 0$ and upward if $V'_G < 0$.

Since the barrier heights are fixed quantities, the movement of the metal Fermi level obviously leads in turn to a distortion in other features of the band diagram. The situation is akin to bending a rubber doll out of shape. Viewed another way, $V'_G \neq 0$ causes potential drops and E_c (E_v) band bending interior to the structure. No band bending occurs, of course, in the metal because it is an equipotential region. In the insulator and semiconductor, however, the energy bands must exhibit an upward slope (increasing E going from the gate toward the back contact) when $V'_G > 0$ and a downward slope when $V'_G < 0$.

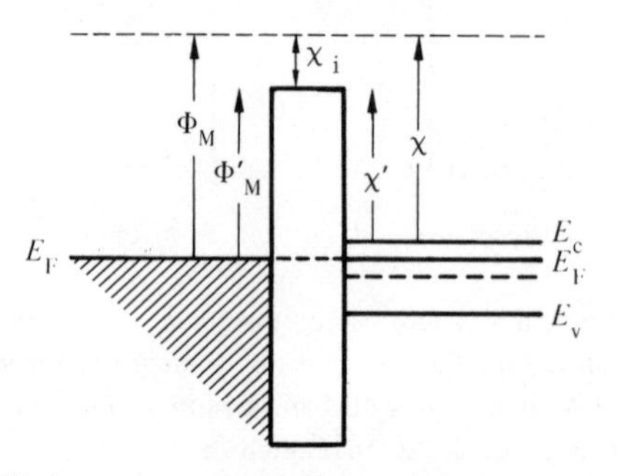

Fig. 2.3 Equilibrium energy band diagram for an ideal MOS structure.

Moreover, the application of Poisson's equation to an ideal insulator with no carriers or charge centers yields $d\mathscr{E}_{\text{oxide}}/dx = 0$ and $\mathscr{E}_{\text{oxide}} =$ constant. Hence, the slope of the energy bands in the ideal insulator is a constant—E_c and E_v are linear functions of position. Naturally, band bending in the semiconductor is expected to be somewhat more complex in its functional form, but, per idealization #5, must always vanish ($\mathscr{E} \to 0$) before reaching the back contact.

2.2.2 The Static State

Given the general principles just discussed, it is now a relatively simple matter to describe the internal status of the ideal MOS structure under various static biasing conditions. Taking the Si substrate to be n-type, consider first the application of a positive bias. The application of $V_G' > 0$ lowers E_F in the metal relative to E_F in the semiconductor and causes a positive sloping of the energy bands in both the insulator and semiconductor. The resulting energy band diagram is shown in Fig. 2.4(a). The major conclusion to be derived from Fig. 2.4(a) is that the electron concentration inside the semiconductor, $n = n_i \exp[(E_F - E_i)/kT]$, increases as one approaches the oxide–semiconductor interface. This particular situation, where the majority carrier concentration is greater near the oxide–semiconductor interface than in the bulk of the semiconductor, is known as *accumulation*.

It is convenient at this point to introduce a second visualization method which is complementary in nature to the energy band diagram and in which one deals solely in terms of the approximate charge distribution inside the MOS structure. From a charge standpoint, the application of $V_G' > 0$ places positive charges on the MOS-C gate. To maintain a balance of charge, negatively charged electrons must be drawn toward the semiconductor–insulator interface—the same conclusion established previously by using the energy band diagram. Thus the charge inside the device as a function of position can be approximated as shown in Fig. 2.4(b). You will note that no attempt has been made to represent the exact charge distribution inside the semiconductor. Rather, a squared-off or block approximation is employed and the resulting figure is called a block charge diagram. Block charge diagrams are intended to be qualitative in nature and the magnitude and spatial extent of the charges should be interpreted with this fact in mind. However, because the electric field in both the metal and semiconductor bulk is zero, the total charge in the structure must also be zero according to Gauss's law. Hence, in constructing block charge diagrams the net positive charge is always drawn equal to the net negative charge.

Consider next the application of a *small* negative potential to the MOS-C gate. The application of a small $V_G' < 0$ slightly raises E_F in the metal relative to E_F in the semiconductor and causes a small negative sloping of the energy bands in both the insulator and semiconductor, as displayed in Fig. 2.4(c). From the diagram it is clear that the concentration of majority carrier electrons has been decreased, depleted, in the vicinity of the oxide–semiconductor interface. A similar conclusion results from charge considerations. Setting $V_G' < 0$ places a minus charge on the gate, which in turn repels electrons from the oxide–semiconductor interface and exposes the positively charged donor sites. The approximate charge distribution is therefore as shown in Fig. 2.4(d). This situation, where the electron and hole concentrations at the oxide–semiconductor interface

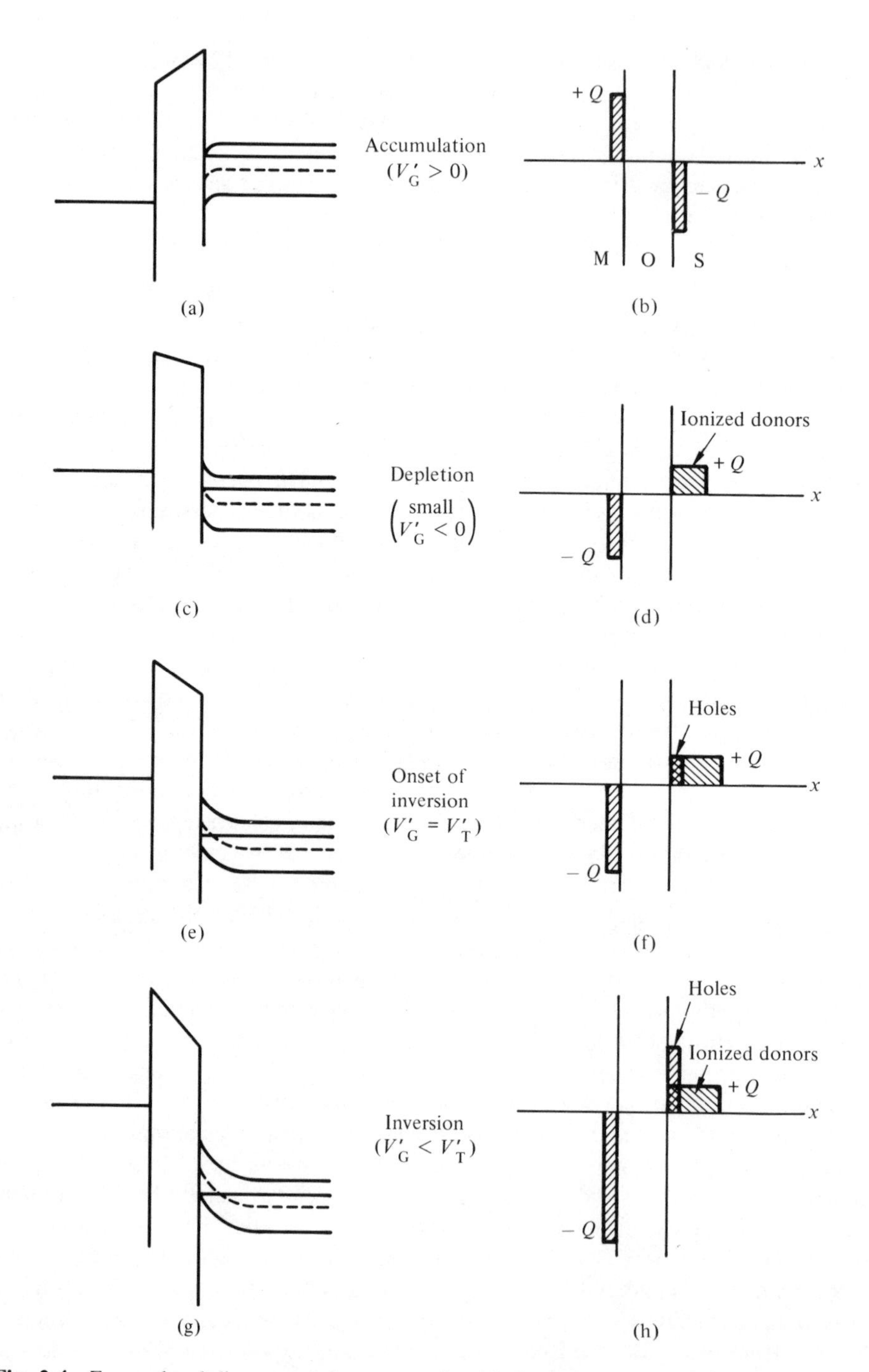

Fig. 2.4 Energy band diagrams and corresponding block charge diagrams describing the static state in an ideal n-type MOS-capacitor.

are less than the background doping concentration (N_A or N_D), is known for obvious reasons as *depletion*.

Finally, suppose a larger and larger negative bias is applied to the MOS-C gate. As V_G' is increased negatively from the situation pictured in Fig. 2.4(c), the bands at the semiconductor surface will bend up more and the hole concentration at the surface (p_s) will likewise increase systematically from less than n_i when E_i (surface) $< E_F$, to n_i when E_i (surface) $= E_F$, to greater than n_i when E_i (surface) exceeds E_F. Eventually, the hole concentration increases to the point shown in Fig. 2.4(e) and (f), where

$$E_i(\text{surface}) - E_i(\text{bulk}) = 2[E_F - E_i(\text{bulk})] \tag{2.2}$$

and

$$p_s = n_i\, e^{E_i(\text{surface})-E_F} = n_i\, e^{E_F-E_i(\text{bulk})} = n_{\text{bulk}} = N_D \tag{2.3}$$

Clearly, when $p_s = N_D$ for the special applied bias $V_G' = V_T'$ the surface is no longer depleted. Moreover, for further increases in negative bias ($V_G' < V_T'$), p_s exceeds $n_{\text{bulk}} = N_D$ and the surface region appears to change in character from n-type to p-type. In accordance with the change in character observation, the $V_G' < V_T'$ situation where the minority carrier concentration at the surface exceeds the bulk majority carrier concentration is referred to as *inversion*. Energy band and block charge diagrams depicting the inversion condition are displayed in Fig. 2.4(g) and (h).

If analogous biasing considerations are performed for an ideal p-type device, the results will be as shown in Fig. 2.5. It is important to note from this figure that biasing regions in a p-type device are reversed in polarity relative to the voltage regions in an n-type device; i.e., accumulation in a p-type device occurs when $V_G' < 0$, etc.

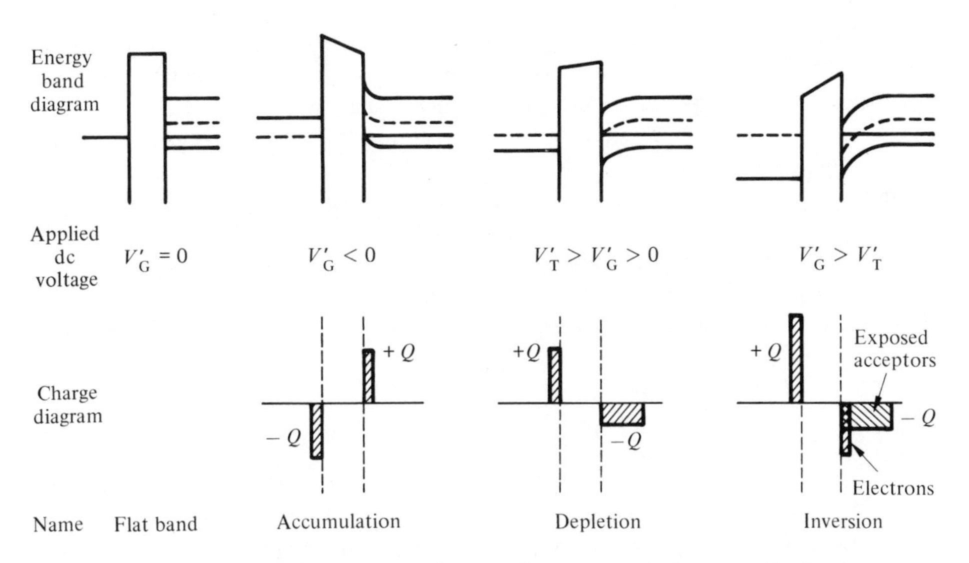

Fig. 2.5 Energy band and block charge diagrams for a p-type device under flat band, accumulation, depletion, and inversion conditions.

In summary, then, one can distinguish three physically distinct biasing regions—accumulation, depletion, and inversion. For an ideal n-type device accumulation occurs when $V_G' > 0$, depletion when $V_T' < V_G' < 0$, and inversion when $V_G' < V_T'$. The cited voltage polarities are simply reversed for an ideal p-type device. No band bending in the semiconductor or *flat band* at $V_G' = 0$ marks the dividing line between accumulation and depletion. The dividing line at $V_G' = V_T'$ is simply called the depletion–inversion transition point, with Eq. (2.2) quantitatively specifying the onset of inversion for both n- and p-type devices.

2.3 SEMICONDUCTOR ELECTROSTATICS

2.3.1 Definition of Parameters

The purpose of this section is to establish analytical relationships for the charge density, electric field, and electrostatic potential existing inside the semiconductor component of an MOS-C under static conditions. Like the *pn* junction analysis (Chapter 2, Volume II), obtaining a mathematical description of the dc state within the semiconductor is tactically a matter of solving Poisson's equation. Although an approximate solution paralleling that presented in the *pn* junction analysis will be included herein, our initial efforts are directed toward obtaining a solution which is "exact" within the ideal structure framework. An exact solution is possible in the MOS-C case because the semiconductor is always in equilibrium regardless of the applied dc bias. As the reader might suspect, however, the exact formulation is somewhat more involved and therefore requires a certain amount of preparatory development.

We begin by letting x be the depth into the semiconductor as measured from the oxide–semiconductor interface (see Fig. 2.6). Note that, under the assumption the semiconductor is sufficiently thick so that the electric field vanishes in the bulk of the material (idealization #5), it is permissible to treat the semiconductor mathematically as if it extended from $x = 0$ to $x = \infty$. Furthermore, since the electrostatic potential is arbitrary to within a constant, we can choose the potential to be zero in the semiconductor bulk; that is, let $V = 0$ at $x = \infty$.

In solving Poisson's equation one could work with the standard system parameters and variables such as the semiconductor doping (N_A, N_D) and the electrostatic potential V. However, in performing mathematical manipulations and in interpreting results, it is far more convenient to deal in terms of normalized parameters. It is therefore customary to introduce the dimensionless quantities,

$$U_F = [E_i(\text{bulk}) - E_F]/kT \tag{2.4}$$

and

$$U = [E_i(\text{bulk}) - E_i(x)]/kT = V/(kT/q) \tag{2.5}$$

U_F and U are also defined graphically in Fig. 2.6. U_F is called the doping parameter and is directly related to the semiconductor doping concentration. $U(x)$ is the electrostatic potential normalized to kT/q and is usually referred to as simply "the potential" if no

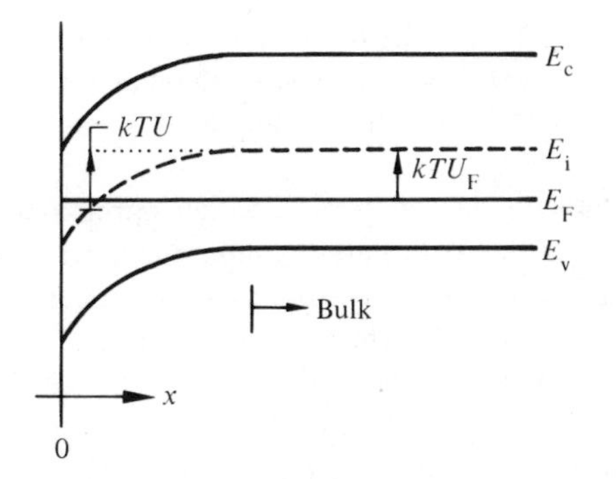

Fig. 2.6 Graphical definition of U and U_F.

ambiguity exists. U evaluated at the oxide-semiconductor interface is given the special symbol, U_S, and is known as the "surface potential." Also note that $U(x \to \infty) = 0$ in agreement with the choice of $V = 0$ at $x = \infty$.

Normalized parameters such as U_F and U are often accepted rather reluctantly. To surmount this problem it is helpful to learn as much as possible about the parameters. Relative to the doping parameter one should know (1) the significance of the sign (plus or minus) associated with U_F, (2) how to calculate U_F for a given impurity concentration, and (3) the range of U_F values normally encountered in practical problems. Since E_F lies above E_i in n-type materials and drops below E_i in p-type materials, an inspection of Eq. (2.4) rapidly reveals $U_F < 0$ given an n-type semiconductor and $U_F > 0$ given a p-type semiconductor. Consequently, the sign of U_F indicates the doping type. Turning next to the computation of U_F, we know from Volume I that

$$n_{\text{bulk}} = n_i e^{[E_F - E_i(\text{bulk})]/kT} = N_D \qquad (\text{if } N_D \gg N_A) \tag{2.6a}$$

$$p_{\text{bulk}} = n_i e^{[E_i(\text{bulk}) - E_F]/kT} = N_A \qquad (\text{if } N_A \gg N_D) \tag{2.6b}$$

Thus, combining Eqs. (2.4) and (2.6) yields

$$U_F = -\ln(N_D/n_i) \qquad \text{for } n\text{-type semiconductor} \tag{2.7a}$$

$$U_F = \ln(N_A/n_i) \qquad \text{for } p\text{-type semiconductor} \tag{2.7b}$$

Finally, semiconductor doping concentrations in MOS devices typically lie somewhere between $10^{14}/\text{cm}^3$ and $10^{17}/\text{cm}^3$. Employing Eqs. (2.7) with $n_i \simeq 10^{10}/\text{cm}^3$ appropriate for Si at room temperature, we therefore find the range of U_F values normally encountered in practical problems is $\boxed{9 \lesssim |U_F| \lesssim 16}$.

The second parameter U is of course a function of both position inside the semiconductor and the voltage applied to the MOS-C gate. Of prime importance is the connection between the U_S value at the oxide–semiconductor interface and the biasing states described in the previous section. Clearly, under flat band conditions $U_S = 0$. Moreover, combining Eqs. (2.2), (2.4), and (2.5), *one concludes* $\boxed{U_S = 2U_F}$ *at the depletion-inversion transition point*. It therefore follows that, in a p-type semiconductor, $U_S < 0$ if the semiconductor is accumulated, $0 < U_S < 2U_F$ if the semiconductor is depleted, and $U_S > 2U_F$ if the semiconductor is inverted. For an n-type semiconductor the inequalities are merely reversed. In other words, knowledge of U_S completely specifies the biasing state inside the semiconductor.

Another item of interest is the range of U_S values normally encountered in practical problems. Although greater excursions are possible, the band bending inside an MOS-C is routinely such that the Fermi level at the oxide-semiconductor interface is confined to band gap energies between E_v and E_c. Thus, given E_v (surface) $\leq E_F \leq E_c$ (surface), and assuming a Si substrate maintained at room temperature, the range of U_S values normally encountered in practice is $\boxed{U_F - 21 \lesssim U_S \lesssim U_F + 21}$. It is left to the reader as an exercise to verify this result. The cited range of U_S values, it should be noted, is also of interest from a theoretical standpoint. When E_F crosses into either the valence band or conduction band at the surface, the surface region becomes decidedly degenerate and degenerate relationships must be employed in calculating the carrier concentrations. Standard quantitative analyses, including the one presented herein, employ nondegenerate relationships and are therefore restricted in validity to the indicated range of U_S values.

In addition to U and U_F, quantitative expressions for the band bending inside of a semiconductor are normally formulated in terms of a special length parameter known as the *intrinsic Debye length*. The Debye length is a characteristic length which was originally introduced in the study of plasmas. (A plasma is a highly ionized gas containing an equal number of positive gas ions and negative electrons.) Whenever a plasma is perturbed by placing a charge in or near it, the mobile species always rearrange so as to shield the plasma proper from the perturbing charge. The Debye length is the shielding distance, or roughly the distance where the electric field emanating from the perturbing charge falls off by a factor of $1/e$. A semiconductor devoid of band bending can be viewed as a type of plasma with its equal number of ionized impurity sites and mobile electrons or holes. The placement of charge near the semiconductor, on the MOS-C gate for example, also causes the mobile species inside the semiconductor to rearrange so as to shield the semiconductor proper from the perturbing charge. The shielding distance or band bending region is again on the order of a Debye length, L_D^*, where

$$L_D^* = \left[\frac{K_S\varepsilon_0 kT}{q^2(n_{bulk} + p_{bulk})}\right]^{1/2} \tag{2.8}$$

Although the Debye length characterization applies only to small deviations from flat band, it is convenient to employ the Debye length appropriate for an *intrinsic* material as a normalizing factor in theoretical expressions. The *intrinsic* Debye length, L_D, is obtained from the more general L_D^* relationship by setting $n_{bulk} = p_{bulk} = n_i$; that is,

$$L_D = \left[\frac{K_S\varepsilon_0 kT}{2q^2 n_i}\right]^{1/2} \tag{2.9}$$

2.3.2 Exact Solution

Expressions for the charge density, electric field, and potential as a function of position inside the semiconductor are obtained by solving Poisson's equation. Since the MOS-C is assumed to be a one-dimensional structure (idealization #7), Poisson's equation simplifies to

$$\frac{d\mathscr{E}}{dx} = \frac{\rho}{K_S\varepsilon_0} = \frac{q}{K_S\varepsilon_0}(p - n + N_D - N_A) \tag{2.10}$$

Maneuvering to recast the equation in a form more amenable to solution, we note

$$\mathscr{E} = \frac{1}{q}\frac{dE_i(x)}{dx} = -\frac{kT}{q}\frac{dU}{dx} \tag{2.11}$$

The first equality in Eq. (2.11) is a restatement of Eq. (3.15) in Volume I. The second equality follows from the Eq. (2.5) definition of U and the fact that $dE_i(\text{bulk})/dx = 0$. In a similar vein we can write

$$p = n_i e^{[E_i(x)-E_F]/kT} = n_i e^{U_F - U(x)} \tag{2.12a}$$

$$n = n_i e^{[E_F-E_i(x)]/kT} = n_i e^{U(x)-U_F} \tag{2.12b}$$

Moreover, since $\rho = 0$ and $U = 0$ in the semiconductor bulk,

$$0 = p_{\text{bulk}} - n_{\text{bulk}} + N_D - N_A = n_i e^{U_F} - n_i e^{-U_F} + N_D - N_A \tag{2.13}$$

or

$$N_D - N_A = n_i(e^{-U_F} - e^{U_F}) \tag{2.14}$$

Substituting the foregoing $\mathscr{E}$, p, n, and $N_D - N_A$ expressions into Eq. (2.10) yields

$$\boxed{\rho = qn_i(e^{U_F-U} - e^{U-U_F} + e^{-U_F} - e^{U_F})} \tag{2.15}$$

and

$$\frac{d^2U}{dx^2} = \left(\frac{q^2 n_i}{K_S\varepsilon_0 kT}\right)(e^{U-U_F} - e^{U_F-U} + e^{U_F} - e^{-U_F}) \tag{2.16}$$

or, in terms of the intrinsic Debye length,

$$\frac{d^2U}{dx^2} = \frac{1}{2L_D^2}(e^{U-U_F} - e^{U_F-U} + e^{U_F} - e^{-U_F}) \tag{2.17}$$

We turn next to the main task at hand. Poisson's equation, Eq. (2.17), is to be solved subject to the boundary conditions:

$$\mathscr{E} = 0 \quad \text{or} \quad \frac{dU}{dx} = 0, \qquad \text{at } x = \infty \tag{2.18a}$$

and

$$U = U_S, \qquad \text{at } x = 0 \tag{2.18b}$$

Multiplying both sides of Eq. (2.17) by dU/dx, integrating from $x = \infty$ to an arbitrary point x, and making use of the Eq. (2.18a) boundary condition, we quickly obtain

$$\mathscr{E}^2 = \left(\frac{kT/q}{L_D}\right)^2 [e^{U_F}(e^{-U} + U - 1) + e^{-U_F}(e^U - U - 1)] \tag{2.19}$$

Equation (2.19) is of the form $y^2 = a^2$, which has two roots, $y = a$ and $y = -a$. As can be deduced by inspection from the energy band diagram, we must have $\mathscr{E} > 0$ when $U > 0$ and $\mathscr{E} < 0$ when $U < 0$. Since the right-hand side of Eq. (2.19) is always positive ($a \geq 0$), the proper polarity for the electric field is obviously obtained by choosing the positive root when $U > 0$ and the negative root when $U < 0$. We can therefore write

$$\mathscr{E} = -\frac{kT}{q}\frac{dU}{dx} = \hat{U}_S \frac{kT}{q}\frac{F(U, U_F)}{L_D} \tag{2.20}$$

where

$$F(U, U_F) \equiv [e^{U_F}(e^{-U} + U - 1) + e^{-U_F}(e^U - U - 1)]^{1/2} \tag{2.21}$$

and

$$\hat{U}_S = \begin{cases} +1 & \text{if } U_S > 0 \\ -1 & \text{if } U_S < 0 \end{cases} \tag{2.22}$$

To complete the solution, one separates the U and x variables in Eq. (2.20) and, making use of the Eq. (2.18b) boundary condition, integrates from $x = 0$ to an arbitrary point x. The end result is Eq. (2.23),

$$\hat{U}_S \int_U^{U_S} \frac{dU'}{F(U', U_F)} = \frac{x}{L_D} \tag{2.23}$$

Although not in a totally explicit form, Eqs. (2.15), (2.20), and (2.23) collectively constitute an exact solution for the electrostatic variables. For a given U_S, numerical techniques can be used to compute U as a function of x from Eq. (2.23). Once U as a function of x is established, direct substitution into Eqs. (2.15) and (2.20) yields ρ and $\mathscr{E}$ as a function of x. Sample plots of U versus x and ρ versus x obtained in the manner just described are presented in Fig. 2.7.

2.3.3 Delta-Depletion Solution

A closed-form though approximate solution for the charge density, electric field, and potential interior to the semiconductor can be established by utilizing the depletion approximation first introduced in the *pn* junction analysis. Because additional approximations based in large part on the nature of the exact solution are also employed in the

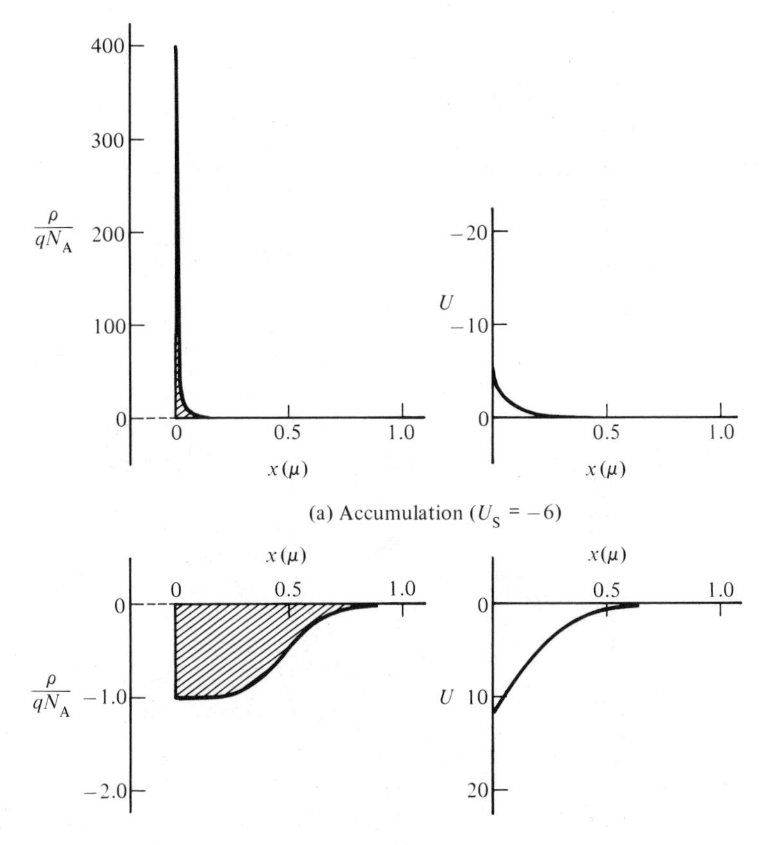

Fig. 2.7 Exact solution for the charge density and potential inside a semiconductor as a function of position assuming $U_F = 12$, $T = 23°C$ and an L_D appropriate for silicon ($L_D = 3.11 \times 10^{-3}$cm). (a) Accumulation ($U_S = 6$) and (b) middle of depletion ($U_S = U_F = 12$). (See next page for continuation of Fig. 2.7(c, d).)

formulation, the approximate relationships for the electrostatic variables are herein referred to collectively as the *delta-depletion solution*.

Since the nature of the exact solution plays a role in the approximate formulation, it is reasonable to spend a few moments examining the plots presented in Fig. 2.7. First of all, note the general correlation between the Fig. 2.7 plots and the semiconductor portion of the diagrams sketched in Fig. 2.5. Next, specifically note that *the charge associated with majority carrier accumulation and minority carrier inversion resides in an extremely narrow portion of the semiconductor immediately adjacent to the oxide–semiconductor interface*. By comparison, the depleted portion of the semicon-

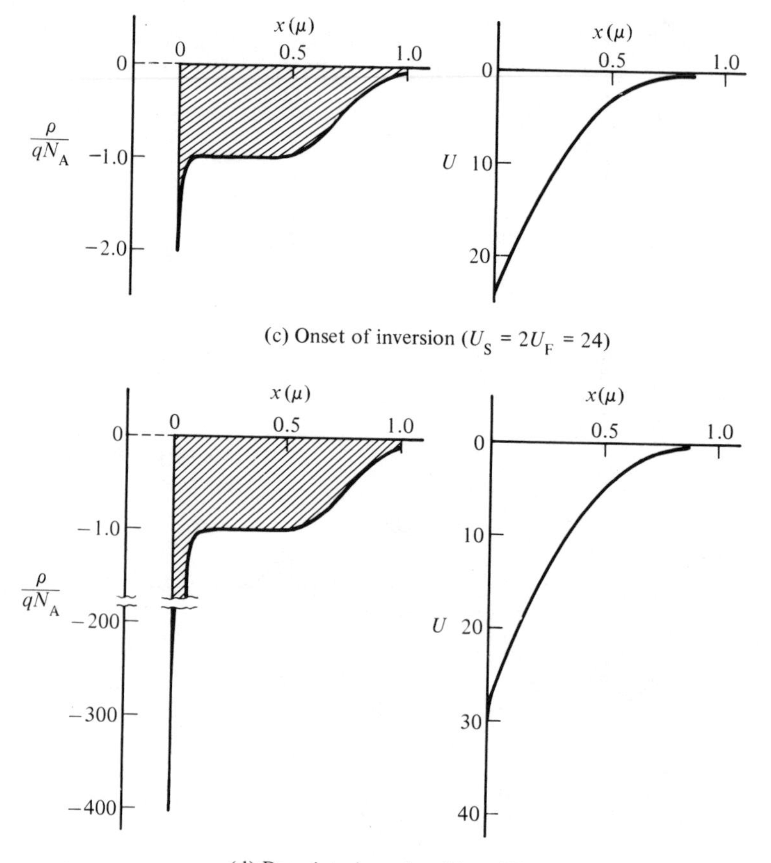

Fig. 2.7 (continued). (c) Onset of inversion ($U_S = 2U_F = 24$), and (d) heavily inverted ($U_S = 2U_F + 6 = 30$). The ρ-diagrams were drawn on a linear scale and the $+U$ axes oriented downward to enhance the correlation with the diagrams sketched in Fig. 2.5.

ductor under moderate depletion biasing extends much deeper into the semiconductor. Moreover, in comparing the depleted semiconductor regions when $U_S = U_F$ (middle of depletion, $U_S = 2U_F$ (onset of inversion), and $U_S = 2U_F + 6$ (inversion), we find *the depletion width increases substantially with increased depletion biasing, but increases only slightly once the semiconductor inverts*. This, we might interject, occurs because the highly peaked inversion charge near the oxide–semiconductor interface is nearly sufficient, in itself, to shield the interior of the semiconductor from any additional charge placed on the MOS-C gate.

In the delta-depletion formulation the very narrow extent of the accumulation and inversion layers is used as justification for approximately representing these layers by

δ-functions of charge located at $x = 0$, the oxide–semiconductor interface. Naturally, invoking the standard depletion approximation means the depleted region existing under depletion and inversion biases is also taken to be terminated abruptly a distance $x = W$ into the semiconductor. However, reflecting the observation that the depletion width increases only slightly once the semiconductor inverts, W is fixed at W_T for all inversion biases, where W_T is the depletion width at the onset of inversion ($U_S = 2U_F$).

The ramifications of the above approximations are as follows: Under accumulation biasing in the delta-depletion formulation the majority carriers pile up in a δ-function distribution at the oxide–semiconductor interface; $\rho = 0$, $\mathscr{E} = 0$, and $U = 0$ for all $x > 0$. In depletion W progressively increases with increased depletion biasing until $W = W_T$; ρ, $\mathscr{E}$, and V for a given depletion bias are computed using the standard depletion approximation. In inversion, minority carriers pile up in a δ-function distribution at the oxide–semiconductor interface, while W, ρ, $\mathscr{E}$, and V for all $x > 0$ remain fixed at their $U_S = 2U_F$ values.

From the foregoing discussion, then, to complete the delta-depletion solution we need only work out expressions for the electrostatic variables when the semiconductor is biased into depletion. Let us perform the required analysis. Invoking the depletion approximation (n and $p \ll N_A$ or N_D), we can write

$$\rho = q(N_D - N_A) \qquad (0 \le x \le W) \tag{2.24}$$

and

$$\frac{d\mathscr{E}}{dx} = -\frac{d^2V}{dx^2} = \frac{q(N_D - N_A)}{K_S\varepsilon_0} \qquad (0 \le x \le W) \tag{2.25}$$

The straightforward integration of Eq. (2.25) employing $\mathscr{E} = 0$ and $V = 0$ at $x = W$ yields

$$\mathscr{E}(x) = \frac{q(N_A - N_D)}{K_S\varepsilon_0}(W - x) \qquad (0 \le x \le W) \tag{2.26}$$

and

$$V(x) = \frac{q(N_A - N_D)}{2K_S\varepsilon_0}(W - x)^2 \qquad (0 \le x \le W) \tag{2.27}$$

The remaining unknown, W, is determined by applying the boundary condition $V = (kT/q)U_S$ at $x = 0$. Thus

$$\frac{kT}{q}U_S = \frac{q(N_A - N_D)}{2K_S\varepsilon_0}W^2 \tag{2.28}$$

and

$$W = \left[\frac{2K_S\varepsilon_0}{q(N_A - N_D)}\frac{kT}{q}U_S\right]^{1/2}. \tag{2.29}$$

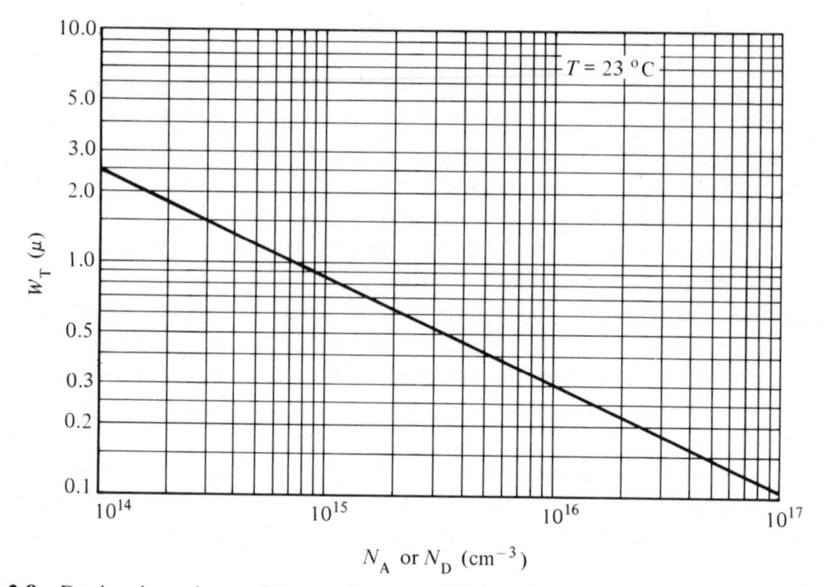

Fig. 2.8 Doping dependence of the maximum equilibrium depletion width inside silicon devices maintained at room temperature.

Since $W = W_T$ when $U_S = 2U_F$, we also conclude

$$W_T = \left[\frac{2K_S\varepsilon_0}{q(N_A - N_D)}\frac{kT}{q}(2U_F)\right]^{1/2} \tag{2.30}$$

A plot of W_T versus doping covering the range $9 \lesssim |U_F| \lesssim 16$ is displayed in Fig. 2.8.

The delta-depletion solution just completed is clearly a relatively gross, first-order theory compared to the exact solution. However, because of its greater simplicity and the direct analogies possible with *pn* junction theory, the approximate formulation is widely employed in the analysis of MOS devices. Quite often the reader will encounter *both* solution approaches in a given problem. An initial analysis based on the delta-depletion formulation usually provides closed-form results that can be readily interpreted. Greater precision is obtained by performing a subsequent analysis based on the exact formulation.

2.4 GATE VOLTAGE RELATIONSHIP

Throughout the discussion of semiconductor electrostatics the biasing state was described in terms of the semiconductor surface potential, U_S. Results formulated in this manner are dependent only on the properties of the semiconductor. U_S, however, is an *internal* system constraint or boundary condition. It is the *externally* applied gate potential, V_G', which is subject to direct control. Thus, if the results of the previous section are to be utilized in practical problems, an expression relating V_G' and U_S must be established. This section is devoted to deriving the required relationship.

We begin by noting that V'_G in the ideal structure is dropped partly across the oxide and partly across the semiconductor or, symbolically,

$$V'_G = \Delta V_{semi} + \Delta V_{ox} \tag{2.31}$$

Because $V = 0$ in the semiconductor bulk, however, the voltage drop across the semiconductor is simply

$$\Delta V_{semi} = V(x = 0) = \frac{kT}{q} U_S \tag{2.32}$$

The task of developing a relationship between V'_G and U_S is therefore reduced to the problem of expressing ΔV_{ox} in terms of U_S.

As stated previously (Section 2.2), in an ideal insulator with no carriers or charge centers

$$\frac{d\mathscr{E}_{ox}}{dx} = 0 \tag{2.33}$$

and

$$\mathscr{E}_{ox} = -\frac{dV_{ox}}{dx} = \text{constant} \tag{2.34}$$

Therefore

$$\Delta V_{ox} = \int_{-x_o}^{0} \mathscr{E}_{ox}\, dx = x_o \mathscr{E}_{ox} \tag{2.35}$$

where x_o is the oxide thickness. The next step is to relate $\mathscr{E}_{ox}$ to the electric field in the semiconductor. The well-known boundary condition on the fields normal to an interface between two dissimilar materials requires

$$(D_{semi} - D_{ox})|_{\text{O-S interface}} = Q_{\text{O-S}} \tag{2.36}$$

where $D = \varepsilon\mathscr{E}$ is the dielectric displacement and $Q_{\text{O-S}}$ is the surface center charge/unit area located at the interface. Since $Q_{\text{O-S}} = 0$ in the idealized structure (idealization #3),*

$$D_{ox} = D_{semi}|_{x=0} \tag{2.37}$$

$$\mathscr{E}_{ox} = \frac{K_S}{K_O} \mathscr{E}_S \tag{2.38}$$

and

$$\Delta V_{ox} = \frac{K_S x_o}{K_O} \mathscr{E}_S = x'_o \mathscr{E}_S \tag{2.39}$$

*The development here is exact. If the delta-depletion formulation is invoked, the δ-function layers of carrier charge at the O-S interface constitute an effective "$Q_{\text{O-S}}$" under accumulation and inversion conditions.

where

$$x_o' = \frac{K_S x_o}{K_O} \tag{2.40}$$

K_S is the semiconductor dielectric constant; K_O, the oxide dielectric constant; and $\mathscr{E}_S$, the electric field in the semiconductor at the oxide–semiconductor interface.

Finally, if Eqs. (2.32) and (2.39) are substituted into Eq. (2.31), we obtain

$$V_G' = \frac{kT}{q} U_S + x_o' \mathscr{E}_S \tag{2.41}$$

or, making use of Eq. (2.20),

$$\boxed{V_G' = \frac{kT}{q}\left[U_S + \hat{U}_S \frac{x_o'}{L_D} F(U_S, U_F)\right]} \tag{2.42}$$

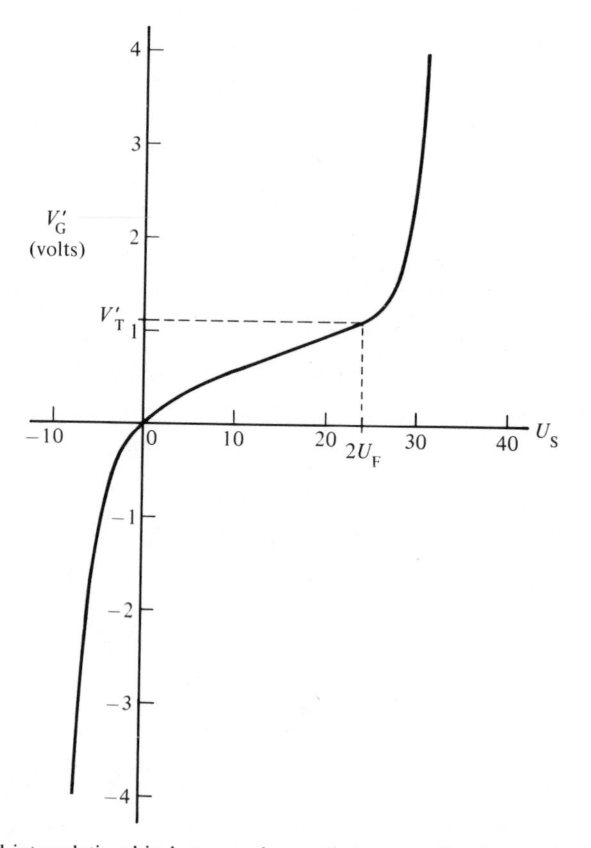

Fig. 2.9 Typical interrelationship between the applied gate voltage and the semiconductor surface potential. ($x_o = 0.1\mu$, $U_F = 12$, $T = 23$°C.)

The $V_G' - U_S$ dependence calculated from Eq. (2.42) employing a typical set of device parameters ($x_o = 0.1\mu$, $U_F = 12$, $T = 23°C$) is displayed in Fig. 2.9. The figure nicely illustrates certain important features of the gate voltage relationship. For one, U_S is a rather rapidly varying function of V_G' when the device is depletion biased. However, when the semiconductor is accumulated ($U_S < 0$) or inverted ($U_S > 2U_F$), it takes a large change in gate voltage to produce a small change in U_S. This implies the gate voltage divides proportionally between the oxide and the semiconductor under depletion biasing. Under accumulation and inversion biasing, on the other hand, changes in the applied potential are dropped almost totally across the oxide. Also note that the depletion bias region extends only from $V_G' = 0$ to $V_G' = 1.10$ V. Since the character of the semiconductor changes drastically in progressing from one side of the depletion bias region to the other, we therefore anticipate a significant variation in the electrical characteristics over a rather narrow range of voltages.

2.5 SUMMARY AND CONCLUDING COMMENTS

The statics of the ideal MOS structure was the main concern of this chapter. Care was taken to clearly define what was envisioned as the ideal structure and, wherever possible, the specific use of a particular idealization was noted. In a subsequent chapter some of the idealizations will be removed and the ensuing perturbations on the device characteristics will be fully examined. The ideal structure, therefore, serves as a reference plane, a foundation for understanding and analyzing the more complex behavior of the real MOS structure.

Qualitatively, MOS statics has been envisioned in terms of the energy band diagram and the block charge diagrams. These visualization aids are, of course, not confined to this single application, but will be utilized again and again in later work. The terms accumulation, flat band, depletion, and inversion were given a physical interpretation using the aforementioned diagrams. Accumulation corresponds to the pile-up of majority carriers at the oxide–semiconductor interface; flat band, to no bending of the semiconductor bands, or equivalently, to no charge in the semiconductor; depletion, to the repulsion of majority carriers from the interface leaving behind an uncompensated impurity-ion charge; and inversion, to the pile-up of minority carriers at the oxide–semiconductor interface. The reader should be able to associate an energy band diagram and block charge diagram with each of these physical situations.

The quantitative formulation of MOS statics is a simple matter of solving Poisson's equation. Two solutions for the semiconductor part of the structure were presented herein: a first-order solution based on the depletion approximation and an exact solution based only on the assumptions of nondegeneracy, a constant doping profile, and a semiconductor of sufficient thickness so that $\mathscr{E} \rightarrow 0$ as $x \rightarrow$ back contact. At first glance the exact solution is complicated by the introduction of the normalized potential, U, and the normalized doping parameter, U_F. However, once these parameters are "digested," the mathematical convenience derived from their use becomes obvious. Knowledge of the parameters general properties greatly aids the "digestion" process. One should specifically remember that, for typical silicon substrates at room temperature, $9 \lesssim |U_F| \lesssim 16$ and $U_F - 21 \lesssim U_S \lesssim U_F + 21$. Furthermore, $U_S = 0$ corresponds to the dividing line between accumulation and depletion while $U_S = 2U_F$ marks the end of depletion and the start of inversion. Finally, it should be remembered that a one-to-one relationship exists between U_S, an internal system constraint, and V_G', the external gate voltage applied to the ideal structure.

PROBLEMS

2.1 For the U_F, U_S parameter sets listed below first indicate the specified biasing condition and then draw the energy band diagram and block charge diagram which characterizes the static state of the system. Assume the MOS structure to be ideal.

(a) $U_F = 12$, $U_S = 12$;

(b) $U_F = -9$, $U_S = 3$;

(c) $U_F = 9$, $U_S = 18$;

(d) $U_F = 15$, $U_S = 36$;

(e) $U_F = -15$, $U_S = 0$.

2.2 Given Si maintained at $T = 23°C$ ($K_S = 11.8$, $kT/q = 0.0255$ V, $n_i = 8.60 \times 10^9/cm^3$) with a donor doping of $N_D = 10^{15}/cm^3$, compute:

(a) L_D

(b) L_D^*

(c) U_F

(d) $\mathscr{E}_S$ when $U_S = U_F$ (exact value)

(e) $\mathscr{E}_S$ when $U_S = 2U_F$ (exact value)

(f) W at $U_S = U_F$

(g) W_T

[*Note*: For more precise work it is necessary to specify the temperature accurately rather than employ the somewhat nebulous term "room temperature." At $T = 23°C$ the most authoritative sources available give an $n_i = 8.60 \times 10^9/cm^3$.]

2.3 In this problem we wish to verify the text statement that the range of U_S values is normally restricted to $U_F - 21 \lesssim U_S \lesssim U_F + 21$ for Si substrates maintained at room temperature.

(a) Referring to Fig. 2.6, draw the semiconductor energy band diagrams corresponding to the situations where (i) $E_F = E_c$ at $x = 0$ and (ii) $E_F = E_v$ at $x = 0$. Indicate kTU_S and kTU_F on your diagrams.

(b) Noting $E_c - E_i \simeq E_i - E_v \simeq E_G/2$, show that:

(i) $U_S \simeq U_F + E_G/2kT$, if $E_F = E_c$ at $x = 0$;
(ii) $U_S \simeq U_F - E_G/2kT$, if $E_F = E_v$ at $x = 0$;
(iii) $U_F - E_G/2kT \lesssim U_S \lesssim U_F + E_G/2kT$, if E_F is confined to band gap energies between E_v and E_c.

(c) At $T = 23°\text{C}$ the Si band gap is 1.12 eV and $kT = 0.0255$ eV. Specialize the (iii) result in part (b) to room temperature ($T \simeq 23°\text{C}$). Is your result compatible with the U_S limits given in the text? (*Note*: The text limits were chosen to be on the "safe" side to allow for variations in the meaning of "room temperature.")

2.4 Let us examine Fig. 2.7, particularly Fig. 2.7(c), more closely.

(a) Draw the block charge diagram describing the charge situation inside an ideal p-bulk MOS-C biased at the onset of inversion.

(b) Is your part (a) diagram in agreement with the plot of ρ/qN_A versus x in Fig. 2.7(c)? Explain why the ρ/qN_A plot has a spike-like nature near $x = 0$ and shows a value of $\rho/qN_A = 2$ at $x = 0$.

(c) Noting that $U_F = 12$ was assumed in constructing Fig. 2.7, determine the W_T predicted by the delta-depletion theory. Is the W_T obtained from the delta-depletion theory consistent with the charge density plot in Fig. 2.7(c)?

2.5 (a) Construct a computer program (possibly employing a hand-held calculator) which will automatically give V_G' versus U_S based on Eq. (2.42) for U_S stepped in single-unit values between the limits $U_F - 21 \leq U_S \leq U_F + 21$. Only U_F and x_o are to be considered input variables. Let $T = 23°\text{C}$ ($kT/q = 0.0255$ V and $L_D = 3.11 \times 10^{-3}$ cm).

(b) Setting $x_o = 0.1\ \mu$, use your program to compute V_G' versus U_S for $U_F = 9$, 12, and 15. Check your $U_F = 12$ results against Fig. 2.9. Note the value of V_T' for each U_F value.

2.6 The exact solution approach presented in Section 2.3 can also be applied to obtain the *exact* electric field and electrostatic potential variation in a pn step junction under *equilibrium* conditions.

(a) Assuming a nondegenerately doped step junction, develop a set of equations that constitutes an exact solution for the electrostatic variables ($\rho, \mathscr{E}, V$) inside the pn junction under equilibrium conditions. In the development you will find it convenient to introduce the normalized potentials: $U_{BI} \equiv \ln(N_A N_D/n_i^2)$; $U_{FP} \equiv \ln(N_A/n_i)$; and $U_{FN} \equiv -\ln(N_D/n_i)$. Set up $x = 0$ at the metallurgical junction, take $U = 0$ at $x = -\infty$ on the p-side of the junction, and let U at $x = 0$ be U_0.

(b) The exact solution approach of Section 2.3 *cannot* be employed to obtain the electrostatic variables inside a pn junction when $V_A \neq 0$. Why is it the solution approach is valid for any applied bias in an MOS-C and not in a pn junction?

3 / Capacitance-Voltage Characteristics

In MOS device analysis the capacitance–voltage (C–V) characteristic is like a picture window, a window revealing the internal nature of the structure. For example, the C–V characteristic provides a means of determining at a glance what band bending is occurring inside the semiconductor for a given applied gate voltage. The C–V characteristic also serves as a powerful diagnostic tool for identifying deviations from the ideal in both the oxide and semiconductor. In fact, a large percentage of the available knowledge about the MOS system was assembled by analyzing the differences between the observed capacitance–voltage characteristics and the MOS-C C–V_G' characteristics predicted on a theoretical basis assuming an ideal structure.

This chapter is devoted primarily to examining the arguments which lead to the expected C–V_G' characteristics of the ideal structure. The majority of the development assumes a static state as described in Chapter 2 and is concerned with two limiting-case solutions—the so-called low-frequency and high-frequency limits. These limiting-case designations refer to the frequency of the ac signal used in the capacitance measurement. In performing the analysis, block charge diagrams are first used to qualitatively correlate the internal behavior of the MOS-C with the general shape of the experimental characteristics. Quantitative low and high frequency theories are next established, employing in turn the delta-depletion and exact-charge formulations. The chapter concludes with clarifying comments about measurement procedures and considerations relevant to a special nonequilibrium condition known as deep depletion.

3.1 QUALITATIVE THEORY

In most laboratories routine C–V measurements (high-frequency measurements) are performed with semiautomatic equipment of the type pictured in Fig. 3.1. The device is positioned on a probing station (normally housed in a light-tight box to exclude room light) and is connected by shielded wires to the C–V bridge. The bridge superimposes a small ac signal, typically less than 15 mV rms at a probing frequency of 1 MHz, on top of a preselected dc voltage, and monitors the resulting ac current flowing into the test structure.

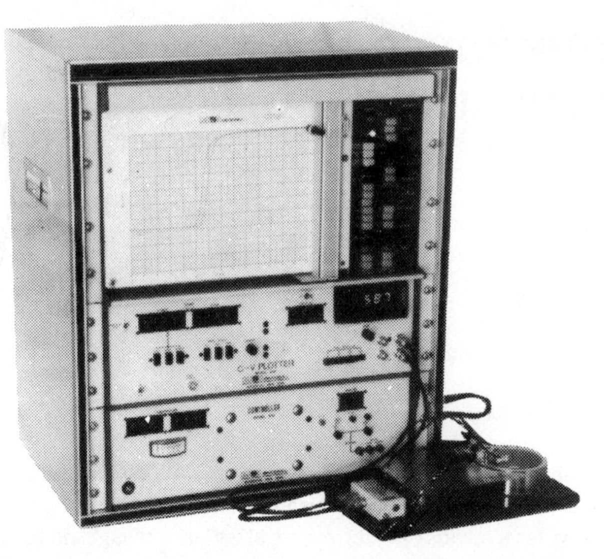

Fig. 3.1 Semiautomatic $C-V$ plotter (Model 868 available from MSI Electronics). The instrument rack contains an $X-Y$ recorder (top), a "smart" 1 MHz $C-V$ bridge (middle), and a temperature controller (bottom). A probing station is also shown attached to the $C-V$ bridge. The $C-V$ bridge has various control options including one for ramping the gate voltage at a preselected rate.

The capacitance determined by the bridge and the dc bias voltage are then fed into a calibrated $X-Y$ recorder. Automatic provisions are usually made for slowly changing the dc voltage to obtain a continuous capacitance versus voltage output on the $X-Y$ recorder. High and low frequency $C-V$ data derived from a representative MOS-capacitor are displayed in Fig. 3.2.

To explain the observed form of the $C-V$ characteristics, let us consider how the charge inside an n-type MOS-C responds to the applied ac signal as the dc bias is systematically changed from accumulation, through depletion, to inversion. We begin with accumulation. In accumulation the dc state is characterized by the pile-up of majority carriers right at the oxide–semiconductor interface. Furthermore, under accumulation conditions the state of the system can be changed very rapidly. For typical semiconductor dopings, the majority carriers, the only carriers involved in the operation of the accumulated device, can equilibrate with a time constant on the order of 10^{-10} to 10^{-13} sec. Consequently, at standard probing frequencies of 1 MHz or less it is reasonable to assume the device can follow the applied ac signal quasi-statically, with the small ac signal adding or subtracting a small ΔQ on the two sides of the oxide as shown in Fig. 3.3(a). Since the ac signal merely adds or subtracts a charge right at the edges of an insulator, the situation inside the accumulated MOS-C is completely analogous to that of an ordinary

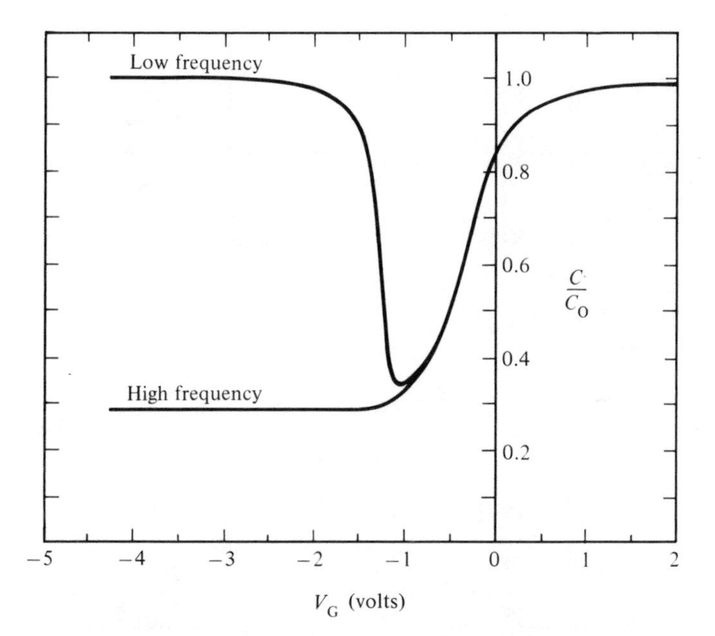

Fig. 3.2 MOS-C high- and low-frequency capacitance–voltage characteristics. The high-frequency curve was derived from a C–V bridge at a measurement frequency of 1 MHz. The low-frequency curve was obtained employing the slow-ramp or quasi-static technique. The device was fabricated on $N_D = 9.1 \times 10^{14}/\text{cm}^3$ (100) Si; $x_o = 0.119\mu$. (Data courtesy of R. R. Razouk, Fairchild Corp.)

parallel-plate capacitor. For either low or high probing frequencies we therefore conclude

$$C(\text{acc}) \simeq C_O = \frac{K_O \varepsilon_0 A_G}{x_o} \tag{3.1}$$

where A_G is the area of the MOS-C gate.

Under depletion biasing the dc state of an n-type MOS structure is characterized by a $-Q$ charge on the gate and a $+Q$ depletion layer charge in the semiconductor. The depletion layer charge is directly related, of course, to the withdrawal of majority carriers from an effective width W adjacent to the oxide–semiconductor interface. Thus, once again, only majority carriers are involved in the operation of the device and the charge state inside the system can be changed very rapidly. As pictured in Fig. 3.3(b), when the ac signal places an increased negative charge on the MOS-C gate, the depletion layer inside the semiconductor widens almost instantaneously; i.e., the depletion width quasi-statically fluctuates about its dc value in response to the applied ac signal. If the stationary dc charge in Fig. 3.3(b) is conceptually eliminated, all that remains is a small fluctuating charge on the two sides of a double-layer insulator. For all probing frequencies this

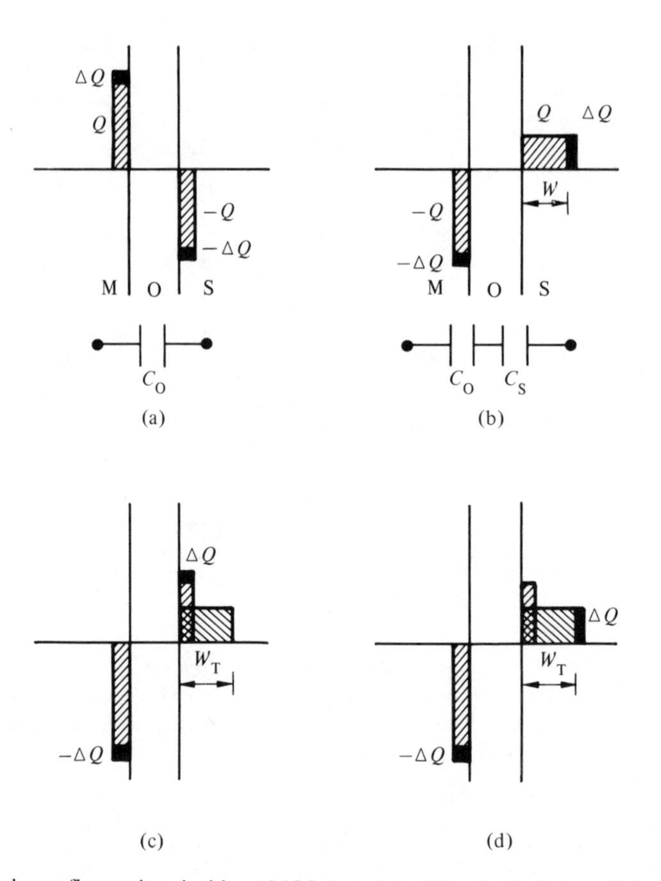

Fig. 3.3 ac charge fluctuations inside an MOS-capacitor under dc biasing conditions corresponding to (a) accumulation, (b) depletion, (c) inversion when $\omega \to 0$, and (d) inversion when $\omega \to \infty$. Equivalent circuit models appropriate for accumulation and depletion biasing are also shown beneath the block charge diagrams in parts (a) and (b), respectively.

situation is clearly analogous to two parallel plate capacitors (C_O and C_S) in series, where

$$C_O = \frac{K_O \varepsilon_0 A_G}{x_o} \qquad \text{(oxide capacitance)} \tag{3.2a}$$

$$C_S = \frac{K_S \varepsilon_0 A_G}{W} \qquad \text{(semiconductor capacitance)} \tag{3.2b}$$

and

$$C(\text{depl}) = \frac{C_O C_S}{C_O + C_S} = \frac{C_O}{1 + W/x_o'} \qquad \left(x_o' \equiv \frac{K_S x_o}{K_O}\right) \tag{3.3}$$

Note from Eq. (3.3) that, because W increases with increased depletion biasing,

C(depl) correspondingly decreases as the dc bias is changed from flat band to the onset of inversion.

Once inversion is achieved we know that an appreciable number of minority carriers pile up near the oxide–semiconductor interface in response to the applied dc bias. Also, the dc width of the depletion layer tends to maximize at W_T. The ac charge response, however, is not immediately obvious. The inversion layer charge might conceivably fluctuate in response to the ac signal as illustrated in Fig. 3.3(c). Alternatively, the semiconductor charge required to balance ΔQ changes in the gate charge might result from small variations in the depletion width as pictured in Fig. 3.3(d). Even a combination of the two extremes is a logical possibility. The problem is to ascertain which alternative describes the actual ac charge fluctuation inside an MOS-C. As it turns out, the observed charge fluctuation depends on the frequency of the ac signal used in the capacitance measurement.

First of all, if the measurement frequency is very low ($\omega \rightarrow 0$), minority carriers can be generated or annihilated in response to the applied ac signal and the time-varying ac state is essentially a succession of dc states. Just as in accumulation, one has a situation (Fig. 3.3(c)) where charge is being added or subtracted right at the edges of a single-layer insulator, and the MOS-C is expected to react like an ordinary capacitor; that is,

$$C(\text{inv}) \simeq C_O \qquad \text{for } \omega \rightarrow 0 \tag{3.4}$$

If, on the other hand, the measurement frequency is very high ($\omega \rightarrow \infty$), the relatively sluggish generation–recombination process will not be able to supply or eliminate minority carriers in response to the applied ac signal. The number of minority carriers in the inversion layer therefore remains fixed at its dc value and the depletion width simply fluctuates about the W_T dc value. Similar to depletion biasing, this situation (Fig. 3.3d) is equivalent to two parallel-plate capacitors in series and

$$C(\text{inv}) = \frac{C_O C_S}{C_O + C_S} = \frac{C_O}{1 + W_T/x_o'} \qquad \text{for } \omega \rightarrow \infty \tag{3.5}$$

W_T is of course a constant independent of the dc inversion bias and $C(\text{inv})_{\omega\rightarrow\infty} = C(\text{depl})_{\text{minimum}}$ = constant for all inversion biases. Finally, if the measurement frequency is such that a *portion* of the inversion layer charge can be created/annihilated in response to the ac signal, an inversion capacitance intermediate between the high and low frequency limits will be observed.

An overall theory can now be constructed by combining the results of the foregoing accumulation, depletion, and inversion considerations. Specifically, we expect the MOS-C capacitance to be approximately constant at C_O under accumulation biases, to decrease as the dc bias progresses through depletion, and to be approximately constant again under inversion biases at a value equal to $\sim C_O$ if $\omega \rightarrow 0$ or $C(\text{depl})_{\text{min}}$ if $\omega \rightarrow \infty$. Moreover, for an n-type device, accumulating gate voltages (where $C \simeq C_O$) are positive, inverting gate voltages are negative, and the decreasing-capacitance, depletion bias region is on the order of a volt or so in width. Quite obviously, this theory for the capacitance–voltage characteristics is in good agreement with the experimental MOS-C C–V_G characteristics presented in Fig. 3.2.

3.2 DELTA-DEPLETION ANALYSIS

A first-order quantitative theory for the low and high frequency C–V_G' characteristics is relatively easy to establish using the delta-depletion formulation. In fact, the qualitative arguments of the previous section need be only slightly modified or can be incorporated directly into the delta-depletion analysis. Relative to modification, the charge blocks representing accumulation and inversion layers in Fig. 3.3 must be replaced by δ-functions of charge located right at the oxide–semiconductor interface. Consequently, the delta-depletion capacitance is *precisely* equal to C_O for accumulation biases and for inversion biases in the low frequency limit. The block charge modeling of the depletion regions in Fig. 3.3, on the other hand, conforms exactly with the simplified charge distributions assumed in the delta-depletion formulation. The depletion and high frequency inversion relationships, Eqs. (3.3) and (3.5), therefore apply without modification. Summarizing,

$$C = \begin{cases} C_O & \text{acc} \qquad (3.6a) \\ \dfrac{C_O}{1 + W/x_o'} & \text{depl} \qquad (3.6b) \\ C_O & \text{inv } (\omega \to 0) \qquad (3.6c) \\ \dfrac{C_O}{1 + W_T/x_o'} & \text{inv } (\omega \to \infty) \qquad (3.6d) \end{cases}$$

Since W_T is a known quantity for a given set of device parameters, Eqs. (3.6) would constitute a complete analytical solution for the C–V_G' characteristics if the depletion bias W were expressed as a function of V_G'. To obtain W as a function of V_G' requires the following manipulations: we know from Chapter 2 that

$$W = \left[\frac{2K_S\varepsilon_0}{q(N_A - N_D)}\frac{kT}{q}U_S\right]^{1/2} \tag{3.7}$$

and

$$V_G' = \frac{kT}{q}U_S + x_o'\mathscr{E}_S \tag{3.8}$$

where, limiting our interests to depletion biases and staying within the framework of the delta-depletion formulation,

$$\mathscr{E}_S = \frac{q(N_A - N_D)}{K_S\varepsilon_0}W \tag{3.9}$$

Substituting Eqs. (3.7) and (3.9) into Eq. (3.8) and rearranging, we obtain

$$W^2 + 2x_o'W - \frac{2K_S\varepsilon_0}{q(N_A - N_D)}V_G' = 0 \tag{3.10}$$

The physically acceptable (i.e., $W \geq 0$) solution to Eq. (3.10) is

$$W = x_o'[(1 + V_G'/V_\delta)^{1/2} - 1] \tag{3.11}$$

giving

$$\boxed{C = \frac{C_O}{\sqrt{1 + V_G'/V_\delta}} \qquad \text{depl}} \tag{3.12}$$

with

$$V_\delta \equiv \frac{q}{2}\frac{K_S x_o^2}{K_O^2 \varepsilon_0}(N_A - N_D) \tag{3.13}$$

Equations (3.6) and (3.12), along with subsidiary relationships for W_T and V_δ, constitute a complete solution for the limiting-case C–V_G' characteristics. A sample set of low and high frequency C–V_G' characteristics constructed using these equations is displayed in Fig. 3.4.

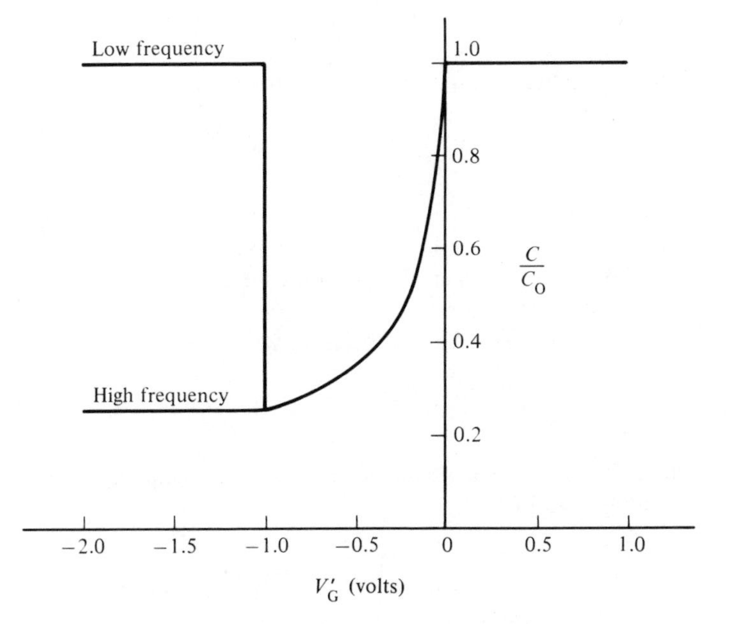

Fig. 3.4 Sample set of low and high frequency C–V_G' characteristics established using the delta-depletion theory. ($x_o = 0.1\mu$, $N_D = 10^{15}/\text{cm}^3$, $T = 23°\text{C}$).

3.3 EXACT CHARGE ANALYSIS

The delta-depletion characteristics, as typified by Fig. 3.4, are a rather crude representation of reality. The first-order theory does a credible job for gate voltages comfortably within a given biasing region, but fails badly in the neighborhood of the transition points going from accumulation to depletion and from depletion to inversion. A more accurate modeling of the observed characteristics is often required in practical applications and is established by working with the exact charge distribution inside the MOS-capacitor. The exact charge analysis for the low frequency capacitance is quite tractable and is reviewed herein. The high frequency analysis, however, is very complex and involved. For this reason the high frequency results are simply quoted and appropriately combined with the low frequency results.

Let us consider an ideal MOS-capacitor with a low frequency ac signal applied to the gate of the device. When the ac gate voltage, v_g', is added to the dc gate voltage, V_G', the charge on the MOS-C gate is of course modified to $Q_G + q_g$, where Q_G and q_g are the dc gate charge per unit area and ac gate charge per unit area, respectively. Provided the device can follow the ac change in gate potential quasi-statically, the assumed case at low operational frequencies, one can state $Q_G(V_G') + q_g$ equals $Q_G(V_G' + v_g')$ or

$$q_g = Q_G(V_G' + v_g') - Q_G(V_G') = \Delta Q_G \tag{3.14}$$

Since, quite generally,

$$C = A_G \frac{q_g}{v_g} \tag{3.15}$$

we have, in the low frequency limit for the ideal structure,

$$C = A_G \frac{q_g}{v_g'} = A_G \frac{\Delta Q_G}{\Delta V_G'} \rightarrow A_G \frac{dQ_G}{dV_G'} \tag{3.16}$$

Equation (3.16) states that the low frequency capacitance can be determined by simply differentiating the dc expression for the gate charge. Instead of working with Q_G directly, it is more convenient to note that in an ideal structure the charge on the gate must balance the charge inside the semiconductor, or $Q_G = -Q_S$, where Q_S is the total semiconductor charge per unit area of the gate. We can therefore write

$$C = -A_G \frac{dQ_S}{dV_G'} = -A_G \frac{dQ_S}{dU_S} \frac{dU_S}{dV_G'} \tag{3.17}$$

The latter form of Eq. (3.17) suggests a mode of attack for completing the analysis. We already know, restating Eq. (2.42),

$$V_G' = \frac{kT}{q}\left[U_S + \hat{U}_S \frac{x_o'}{L_D} F(U_S, U_F)\right] \tag{3.18}$$

In addition, applying Gauss' law, we find that

$$Q_S = -K_S\varepsilon_0\mathscr{E}_S = -\hat{U}_S \frac{kT}{q} \frac{K_S\varepsilon_0}{L_D} F(U_S, U_F) \tag{3.19}$$

Thus, performing the required differentiations, substituting into Eq. (3.17), and reorganizing the result, we conclude

$$C = \frac{C_O}{1 + W_{eff}/x_o'} \tag{3.20}$$

$$W_{eff} = \hat{U}_S L_D \left[\frac{2F(U_S, U_F)}{e^{U_F}(1 - e^{-U_S}) + e^{-U_F}(e^{U_S} - 1)/(1 + \Delta)} \right] \tag{3.21}$$

where

$$\Delta = 0 \qquad \text{in the low frequency limit} \tag{3.22}$$

and, in the high-frequency limit, for a p-type device,

$$\Delta = \begin{cases} 0 & \text{acc} \quad (U_S < 0,\ U_F > 0) \quad (3.23a) \\ \dfrac{(e^{U_S} - U_S - 1)/F(U_S, U_F)}{\displaystyle\int_0^{U_S} \frac{e^{U_F}(1 - e^{-U})(e^{U} - U - 1)}{2F^3(U, U_F)}\, dU} & \begin{array}{l}\text{depl, inv} \\ (U_S > 0,\ U_F > 0)\end{array} \quad (3.23b) \end{cases}$$

Unlike the delta-depletion result, C cannot be expressed explicitly as a function of V_G' in the exact charge formulation. Both variables, however, have been related to U_S and it is possible to compute numerically the capacitance expected from the structure for a given applied gate voltage using Eqs. (3.20) through (3.23) in conjunction with Eq. (3.18). The usual and most efficient computational procedure is to calculate C and the corresponding V_G' for a set of assumed U_S values. Typically, a sufficient set of (C, V_G') points to construct the C–V_G' characteristic will be generated if U_S is stepped by whole-number units $(-5, -4, \cdots)$ over the normal operating range of U_S values ($U_F - 21 \lesssim U_S \lesssim U_F + 21$ at room temperature). It should be noted that care must be exercised if $U_S = 0$ is included as one of the computational points. At $U_S = 0$ the Eq. (3.21) expression for W_{eff} becomes indeterminate (0/0) and, as is readily established, must be replaced by $W_{eff} = \sqrt{2}\, L_D/[\exp(U_F) + \exp(-U_F)]^{1/2}$. Also, the quoted high frequency results hold only for p-type devices. It is nevertheless possible to obtain an n-type characteristic by simply running the calculations for an equivalently doped p-type device and then changing the sign of all computed V_G' values. This procedure works because of the voltage symmetry between ideal n- and p-type devices.

A number of sample C–V_G' characteristics are displayed in Figs. 3.5 to 3.7. These figures exhibit the general doping (Fig. 3.5), oxide thickness (Fig. 3.6), and temperature (Fig. 3.7) dependences of the capacitance–voltage relationship. Note in particular from Fig. 3.5 the significant increase in the high-frequency inversion capacitance and the substantial widening of the depletion bias region with increased doping. In fact, at very high dopings (not shown) the capacitance approaches a constant independent of bias. This should not be an unexpected result, for with increased doping the semiconductor begins to look more and more like a metal and the MOS structure should be expected to react more and more like a standard capacitor. As illustrated in Fig. 3.6, an increase in the oxide

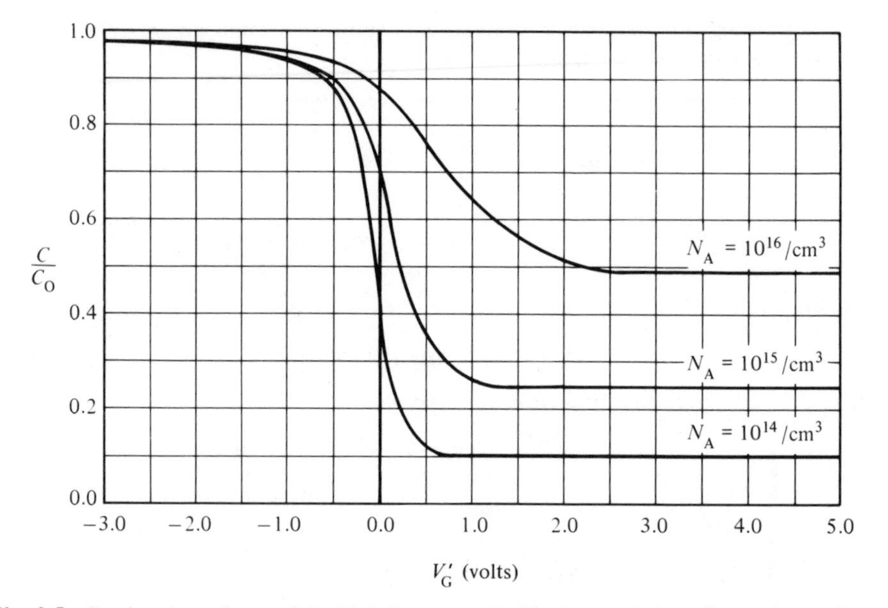

Fig. 3.5 Doping dependence of the high frequency C–V'_G characteristics. (Exact charge theory, $x_o = 0.1\mu$, $T = 23°C$.)

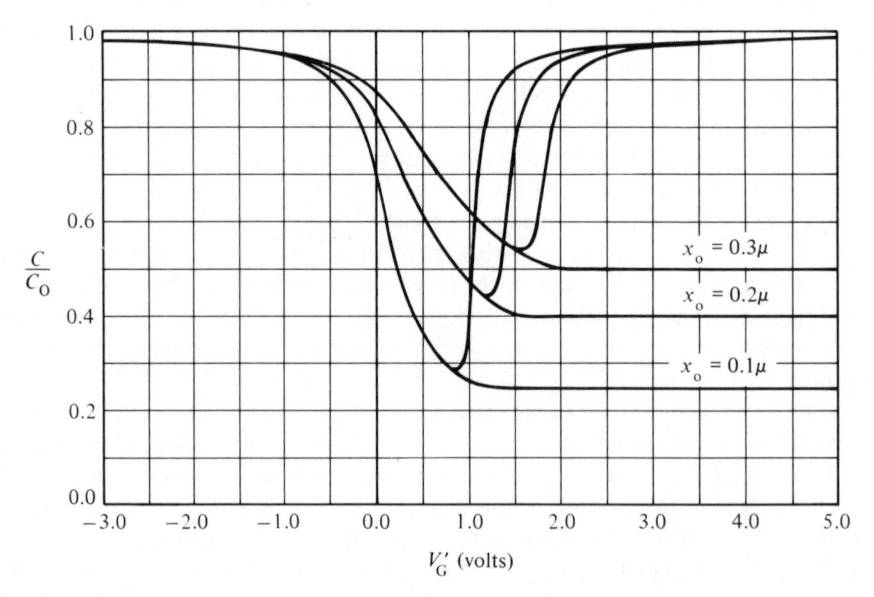

Fig. 3.6 Oxide thickness dependence of the low and high frequency C–V'_G characteristics. (Exact charge theory, $N_A = 10^{15}/cm^3$, $T = 23°C$.)

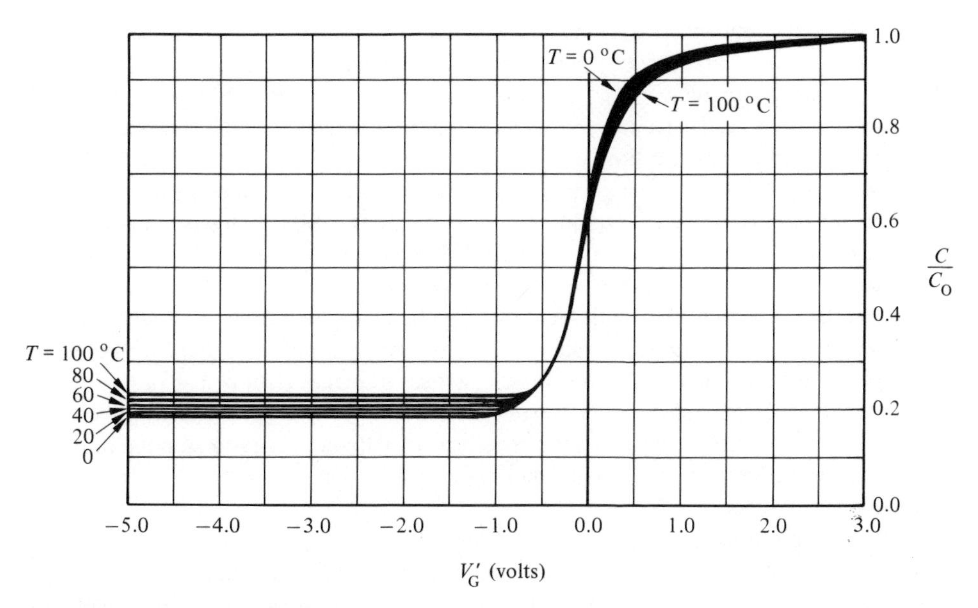

Fig. 3.7 Temperature dependence of the high frequency C–V'_G characteristics. (Exact charge theory, $x_o = 0.1\mu$, $N_D = 5 \times 10^{14}/\text{cm}^3$.)

thickness also widens the depletion bias region and affects the high-frequency inversion capacitance. The increased width of the depletion bias region with increased x_o is simply a consequence of a proportionate increase in the voltage drop across the oxide component of the structure. Finally, Fig. 3.7 nicely displays the moderate sensitivity of the inversion bias capacitance and insensitivity of the ideal structure, depletion bias capacitance to changes in temperature.

3.4 PRACTICAL CONSIDERATIONS/DEEP DEPLETION

In the discussion so far we have more or less sidestepped any clarification of precisely what was meant by "low frequency" and "high frequency" in terms of actual measurement frequencies. One might wonder, will a 100 Hz ac signal typically yield low-frequency C–V characteristics? Perhaps surprisingly, the answer to the question is *no*. Given modern-day MOS-C's with their long carrier lifetimes and low carrier generation rates, even probing frequencies as low as 10 Hz, the practical limit in bridge-type measurements, will yield high-frequency type characteristics. If an MOS-C low-frequency characteristic is required, indirect means such as the quasi-static technique* must be employed to construct the characteristic.

*See M. Kuhn, A quasi-static technique for MOS C–V and surface state measurements, *Solid-State Electronics* 13 (1970):873.

On the high-frequency side, one cannot actually let $\omega \rightarrow \infty$ and expect to observe a high-frequency characteristic. Measurement frequencies, in fact, seldom exceed 1 MHz. At higher frequencies the resistance of the semiconductor bulk comes into play and lowers the observed capacitance. At even high frequencies ($\gtrsim$1 GHz) one must begin to worry about the response time of the majority carriers.

In summary, then, it is the high-frequency characteristic that is routinely recorded and the standard, almost universal measurement frequency is 1 MHz. This is not to say the high-frequency characteristics can be recorded without exercising a certain amount of caution. Suppose, for example, the C–V measurement is performed precisely as described earlier in this chapter, with the dc voltage being ramped from accumulation into inversion to obtain a continuous capacitance versus voltage output on the X–Y recorder. Figure 3.8 illustrates the usual results of such a measurement performed at various ramp rates. Note that at even the slowest ramp rates one does not properly plot out the inversion portion of the high-frequency characteristic. It is actually necessary to stop the ramp in inversion and allow the device to equilibrate before measuring the high-frequency inversion capacitance.

The discussion in the preceding paragraph really serves two purposes, the second being a lead into the important topic of deep depletion. Let us examine the ramped measurement in greater detail. When the ramp voltage is in accumulation or depletion, only majority carriers are involved in the operation of the device, and the dc charge configuration inside the structure rapidly reacts to the changing gate bias. (If the semiconductor can quasi-statically follow a 1 MHz ac signal, it can certainly follow a rather slowly varying voltage ramp.) As the ramp progresses into the inversion bias region, however, a significant number of minority carriers are required to attain an equilibrium charge distribution within the MOS-C. The minority carriers were not present prior to entering the inversion bias region, cannot enter the semiconductor from the remote back contact or across the oxide, and therefore must be created in the near-surface region of the semiconductor. The generation process, as we have noted several times, is rather sluggish and has difficulty supplying the minority carriers needed for the structure to equilibrate. Thus, as pictured in Fig. 3.9(a), the semiconductor is driven into a *nonequilibrium* condition where, in balancing the charge added to the MOS-C gate, the depletion width becomes greater than W_T to offset the missing minority carriers. The described condition, the nonequilibrium condition where there is a deficit of minority carriers and a depletion width in excess of the equilibrium value, is referred to as *deep depletion*.

The existence of a $W > W_T$ of course explains the reduced values of C observed in the ramp measurement. Moreover, the decrease in capacitance with increased ramp rate indicates a greater deficit of minority carriers and a wider depletion width. This is logical, since the greater the ramp rate, the fewer the number of minority carriers generated prior to arriving at a given inversion bias.

The limiting case as far as deep depletion is concerned occurs when the semiconductor is totally devoid of minority carriers—totally deep depleted. Except for a wider depletion width, the total deep depletion condition shown in Fig. 3.9(b) is precisely the same as

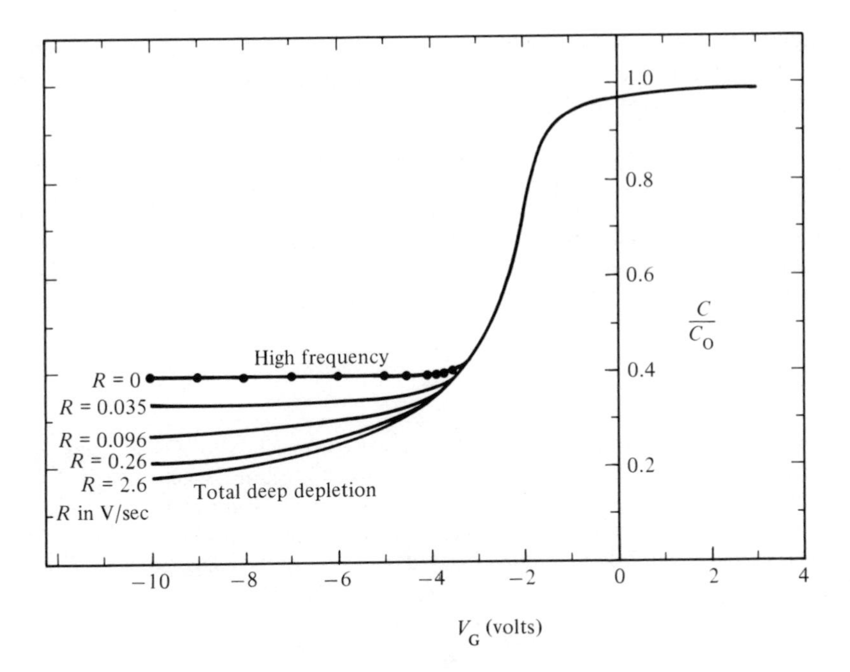

Fig. 3.8 Measured C–V characteristics as a function of the ramp rate (R). In inversion the high-frequency capacitance was obtained by stopping the ramp and allowing the device to equilibrate.

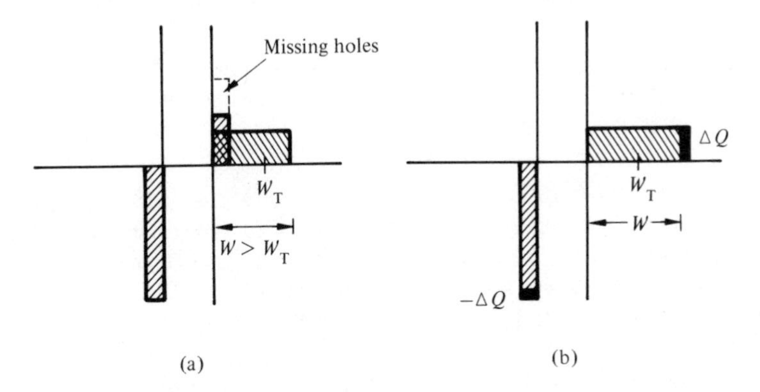

Fig. 3.9 (a) Nonequilibrium charge configuration inside an MOS-capacitor under deep depletion conditions. (b) ac charge fluctuations inside the MOS-C when the semiconductor is totally deep depleted.

the simple depletion condition pictured in Fig. 3.3(b). Consequently, by analogy, and based on the delta-depletion formulation, the limiting-case capacitance exhibited by the structure under deep depletion conditions should be

$$C = \frac{C_O}{\sqrt{1 + V_G'/V_\delta}} \quad \begin{array}{l}\text{total deep depletion} \\ (V_G' > V_T' \ p\text{-type};\ V_G' < V_T' \ n\text{-type.})\end{array} \tag{3.24}$$

Equation (3.24) is in excellent agreement with experimental observations and is essentially identical to the result obtained from an exact charge analysis. The 2.6 V/sec ramp rate curve shown in Fig. 3.8 is an example of a (total) deep depletion characteristic.

PROBLEMS

3.1 With modern-day processing it is possible to produce Semiconductor–Oxide–Semiconductor (SOS) capacitors in which a semiconductor replaces the metallic gate in a standard MOS-C. Answer the questions posed below assuming an SOS-C composed of two *identical n-type* nondegenerate silicon electrodes, an *ideal structure*, and a biasing arrangement as defined by Fig. P3.1. Include any comments which will help to forestall a misinterpretation of the requested pictorial answers.

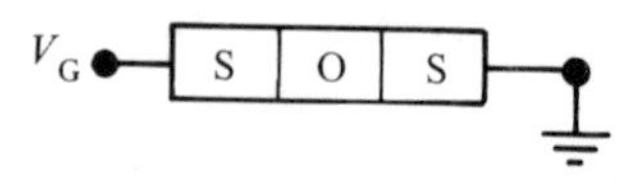

Figure P3.1

(a) Draw the energy band diagrams for the structure when (i) $V_G = 0$, (ii) $V_G > 0$ but small, (iii) $V_G > 0$ and very large, (iv) $V_G < 0$ but small, and (v) $V_G < 0$ and very large.

(b) Draw the block charge diagrams corresponding to the five biasing conditions considered in part (a).

(c) Sketch the expected shape of the high frequency C–V_G characteristic for the SOS-C described in this problem. For reference purposes, also sketch on the same plot the high frequency C–V_G characteristic of an MOS-C assumed to have the same semiconductor doping and oxide thickness as the SOS-C.

Practical notes: (1) Silicon is now routinely employed as the gate material in many "M"OS structures. However, the Si-gate is heavily doped and polycrystalline in nature. A "SOS" structure of the type considered in this problem might be produced by laser annealing an undoped polycrystalline gate, but there is no practical use for such a structure. (2) The acronym SOS found in the device literature stands for Silicon–on–Sapphire and not Semiconductor–Oxide–Semiconductor.

3.2 The experimental C–V_G characteristic shown in Fig. P3.2 was observed under the following conditions: The dc bias was changed very slowly from point (1) to point (2). At point (2) the V_G sweep rate was increased substantially. Upon arriving at point (3) the sweep was stopped for a period of time during which the capacitance decayed to (3′). Finally, the bias was moved very slowly back to (2). Qualitatively explain the observed characteristic.

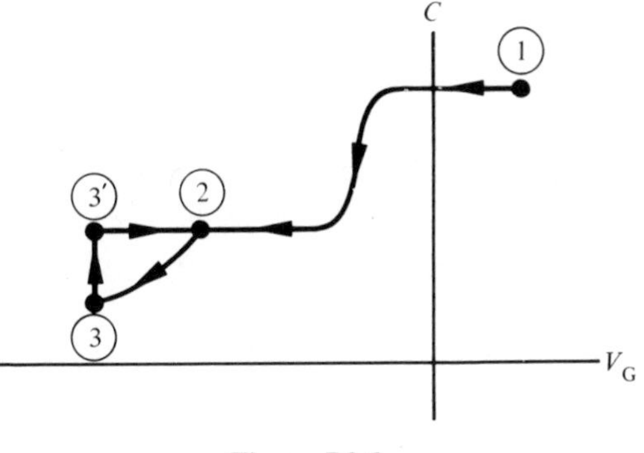

Figure P3.2

3.3 Invoking the delta-depletion approximation and assuming an ideal MOS-C with $x_o = 0.2\ \mu$, $N_D = 10^{15}/\text{cm}^3$, $K_O = 3.9$, $K_S = 11.8$ and $T = 23°\text{C}$,

(a) Compute V_δ

(b) Compute W_T

(c) Compute V_T'

(d) Using the delta-depletion theory, compute C/C_O vs. V_G' for V_G' values at 0.2 V intervals between $V_G' = 0$ and $V_G' = V_T'$. Plot the computed C/C_O–V_G' points on graph paper. Also draw in the accumulation and inversion (high frequency) portions of the characteristic.

(e) Indicate on your plot the characteristic that would be observed if the semiconductor were totally deep depleted. (Compute a sufficient number of C/C_O–V_G' points so that a smooth curve may be constructed out to $V_G' = -5$ V.)

(f) The parameters assumed in this problem are very similar to the parameters of the device yielding the C–V_G characteristics displayed in Fig. 3.8. Compare your curves to the corresponding curves in Fig. 3.8 and comment on the comparison.

3.4 Verify the exact charge result for the *low-frequency* capacitance by supplying the missing mathematical steps in the derivation of Eqs. (3.20) to (3.22).

3.5 (a) Write a computer program that can be used to construct the *low-frequency* C/C_O versus V_G' characteristics based on the exact charge formulation. The program is to calculate C/C_O and the corresponding V_G' for whole number U_S values over the range $U_F - 21 \le U_S \le U_F + 21$. Because of the indeterminate nature of W_{eff} at $U_S = 0$, take special precautions in calculating the flat band capacitance. Let $T = 23°\text{C}$, $kT/q = 0.0255$ V, $L_D = 3.11 \times 10^{-3}$ cm, $K_S = 11.8$, and $K_O = 3.9$. Only U_F and x_o are to be considered input variables.

(b) Setting $U_F = 11.66$ ($N_A = 10^{15}/\text{cm}^3$), use your program to compute C/C_O versus V_G' for $x_o = 0.1\ \mu$, $0.2\ \mu$, and $0.3\ \mu$. Compare your program results with Fig. 3.6.

3.6 The x_o and U_F parameters required in constructing the theoretical C–V_G' characteristic to be compared with a given experimental characteristic are often deduced directly from the experimental C–V_G data. Let us explore the x_o and U_F determination procedures.

(a) The device yielding the high-frequency C–V_G characteristic shown in Fig. 3.8 exhibited a maximum capacitance (C_O) of 82.0 pf. The gate area of the MOS-C was equal to 4.75×10^{-3} cm^2 and $K_O = 3.9$. Determine x_o from the given data.

(b) From the value of C/C_O under strong inversion biasing ($V_G \lesssim -4$ V in Fig. 3.8) calculate W_{eff} utilizing Eq. (3.20), $K_S = 11.8$, and the x_o determined in part (a). Assuming $L_D = 3.11 \times 10^{-3}$ cm, record the value of W_{eff} (strong inversion)$/L_D$.

(c) U. Hartmann and P. Schley, *Phys. Stat. Sol.* (a) 42 (1977):667, established that

$$\frac{W_{eff}(\text{strong inv})}{L_D} = 2e^{-|U_F|/2}\left[17.28 + 99.48 \tanh\left(\frac{|U_F| - 7.936}{47.13}\right)\right]^{1/2}$$

to an accuracy of better than $\pm$ 0.1% over the range $4 \leq |U_F| \leq 16$. By making a plot of W_{eff} (strong inv)$/L_D$ versus U_F, through iterative techniques using a computer, or by a simple hit-and-miss method, determine the value of U_F for the Fig. 3.8 device to four significant figures. (*Note*: A rather good estimate of N_D and therefore U_F can be rapidly deduced from Fig. 2.8 by equating W_{eff} (strong inv) and W_T.)

4 / Deviations from the MOS Ideal

Deviations from the ideal, the norm, the expected are what make life interesting and challenging. A similar statement can be made concerning MOS devices. Although the ideal structure is excellent for establishing the basic principles of MOS theory in a clear and uncomplicated fashion, the scientific community would have lost interest in an ideal structure long ago. Indeed, a large portion of the research and routine characterization of MOS devices was and continues to be concerned with understanding, modeling and minimizing deviations from the ideal. In this chapter we examine four of the more important and most commonly encountered deviations from the ideal; namely, a nonzero metal–semiconductor workfunction difference, mobile ions in the oxide, the built-in or fixed oxide charge, and interfacial traps. In each case the net effect of the nonideality is identified and correlated with its perturbation on the observed C–V characteristics. Because of their practical importance, all of the topics to be discussed have been extensively researched and the mass of available information is quite imposing. The reader should not be surprised, however, to find some unanswered or only partially answered questions about a given nonideality.

4.1 METAL–SEMICONDUCTOR WORKFUNCTION DIFFERENCE

The energy band diagrams for the isolated components of an Al–SiO_2–(p-type) Si system are drawn roughly to scale in Fig. 4.1(a). Upon examining this figure we see that in a real device the energy difference between the Fermi energy and the vacuum level is unlikely to be the same in the isolated metal and semiconductor components of the system; that is, in contrast to the ideal structure, $\Phi_M \neq \chi + (E_c - E_F)_\infty$. To correctly describe real systems the ideal theory must be modified to account for this metal–semiconductor workfunction difference.

In working toward the required modification, let us first construct the equilibrium ($V_G = 0$) energy band diagram appropriate for the system. We begin by conceptually connecting a wire between the outer ends of the metal and semiconductor shown in Fig. 4.1(a) and bringing the two materials together in a vacuum until they are a

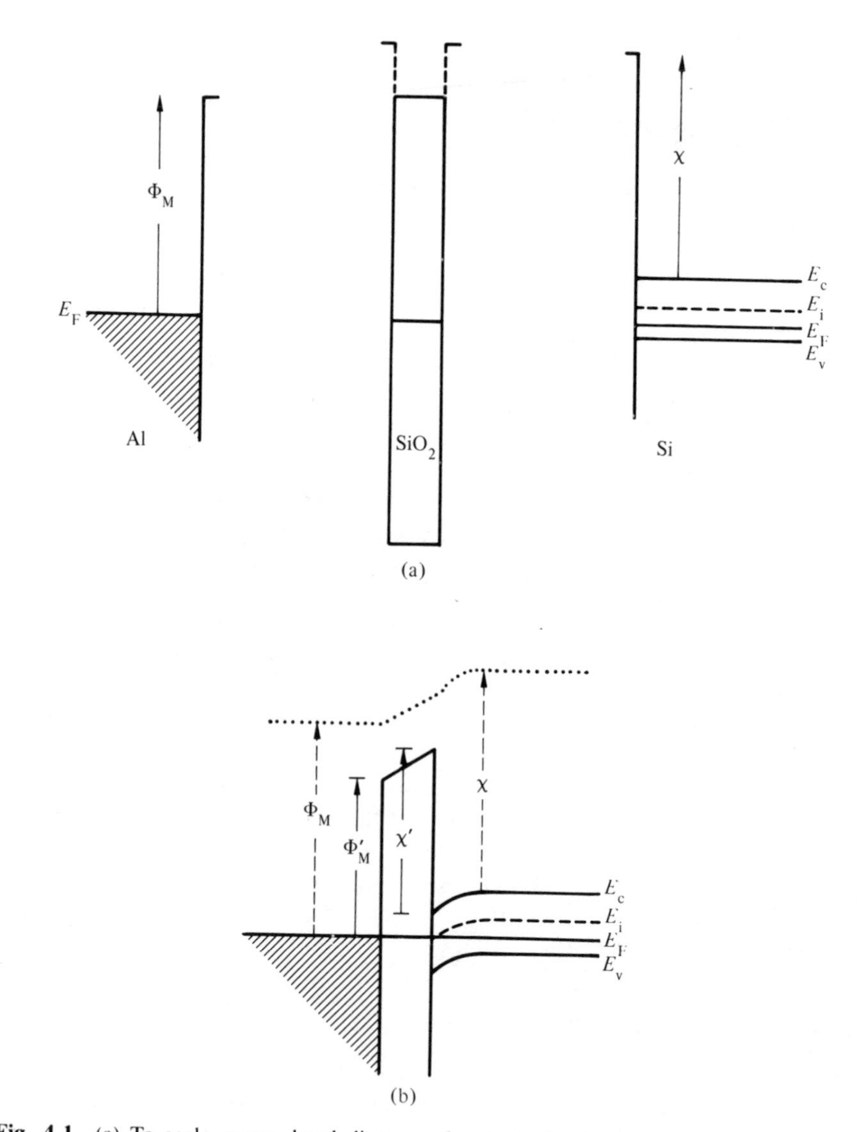

Fig. 4.1 (a) To scale energy band diagrams for the isolated components of the Al–SiO_2–Si system. (b) Equilibrium ($V_G = 0$) energy band diagram typical of real MOS structures.

distance x_o apart. The connecting wire facilitates the transfer of charge between the metal and semiconductor and helps maintain the system in an equilibrium state where the respective Fermi levels "line up" as the materials are brought together. With the metal E_F and semiconductor E_F at the same energy, and $\Phi_M \neq \chi + (E_C - E_F)_\infty$, the vacuum levels in the two materials must clearly be dissimilar. Thus, an electric field, $\mathscr{E}_{vac}$, develops between the components (with the Si vacuum level above the Al vacuum level given the situation pictured in Fig. 4.1a) and band bending occurs inside the semiconductor ($K_S\mathscr{E}_S$ must equal $\mathscr{E}_{vac}$); $\mathscr{E}_{vac}$ and the semiconductor band bending increase of course as the components are brought closer and closer together. Once the metal and semiconductor are positioned a distance x_o apart, the insulator is next inserted into the empty space between the other two components. The addition of the insulator simply lowers the effective surface barriers ($\Phi_M \rightarrow \Phi_M'$ and $\chi \rightarrow \chi'$) and reduces the electric field in the x_o region ($K_O > 1$). The resulting equilibrium energy band diagram typical of real MOS systems is shown in Fig. 4.1(b).

The point to be derived from the preceding argument and Fig. 4.1(b) is that the workfunction difference, and any deviation from the ideal for that matter, modifies the relationship between the semiconductor surface potential and the applied gate voltage. $V_G = 0$, for example, does not give rise to flat band conditions inside the semiconductor. To establish the new V_G–U_S relationship, suppose an arbitrary gate voltage is applied to the Fig. 4.1(b) structure, yielding the situation pictured in Fig. 4.2(a). Equating the energies from the semiconductor Fermi-level to the top of the band diagram as viewed from the two sides of the insulator in Fig. 4.2(a), we obtain

$$\underbrace{\Phi_M' - qV_G + q\Delta V_{ox}}_{\text{metal side}} = \underbrace{(E_c - E_F)_\infty - kTU_S + \chi'}_{\text{semiconductor side}} \tag{4.1a}$$

or

$$V_G - \left(\frac{kT}{q}U_S + \Delta V_{ox}\right) = \frac{1}{q}[\Phi_M' - \chi' - (E_c - E_F)_\infty] \tag{4.1b}$$

Introducing ϕ_{MS}, the metal–semiconductor workfunction difference (in volts), where

$$\phi_{MS} \equiv \frac{1}{q}[\Phi_M' - \chi' - (E_c - E_F)_\infty] \tag{4.2a}$$

$$= \frac{1}{q}[\Phi_M - \chi - (E_c - E_F)_\infty] \tag{4.2b}$$

and recalling

$$V_G' = \frac{kT}{q}U_S + \Delta V_{ox} \tag{4.3}$$

we conclude

$$\boxed{\Delta V_G \equiv (V_G - V_G')|_{\text{same } U_S} = \phi_{MS}} \tag{4.4}$$

Equation (4.4) is interpreted as follows: to achieve a given U_S value inside a nonideal structure, one must apply a gate voltage equal to $V_G' + \phi_{MS}$, where V_G' is the gate voltage required to achieve the same U_S value if the structure were ideal. For example, suppose $\phi_{MS} = -1$ V and flat band conditions are desired. In the ideal structure, flat band occurs at $V_G' = 0$. Thus in the $\phi_{MS} = -1$ V structure, flat band occurs at $V_G' + \phi_{MS} = -1$ V.

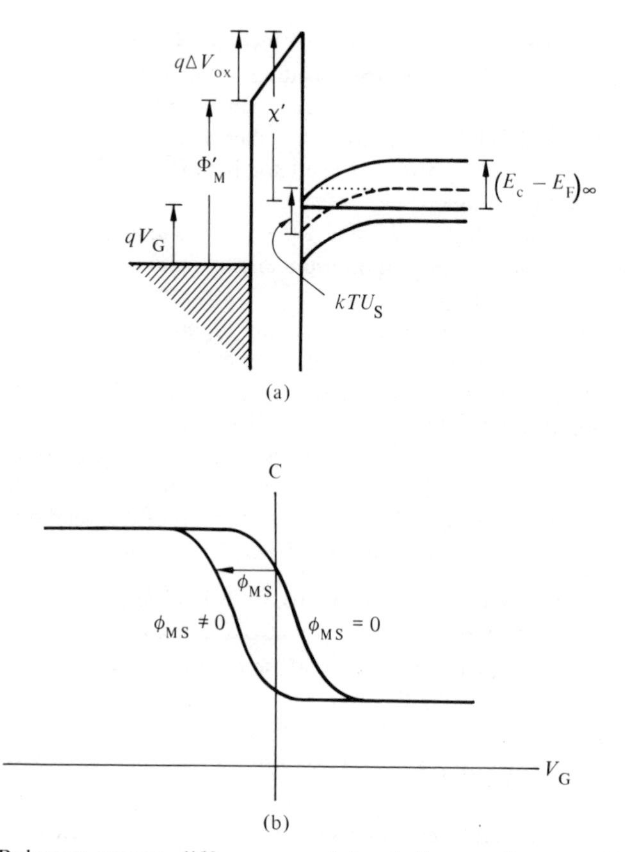

Fig. 4.2 (a) Relevant energy differences and band diagram for an arbitrarily biased $\phi_{MS} \neq 0$ MOS structure. (b) Effect of a $\phi_{MS} \neq 0$ on the MOS-C high-frequency C–V characteristics.

Likewise, if the transition point voltage in the ideal structure occurs at $V_T' = 1.5$ V, the transition point voltage in the $\phi_{MS} = -1$ V version of the structure will occur at $V_T = V_T' + \phi_{MS} = 0.5$ V. Relative to the capacitance–voltage characteristics, since ϕ_{MS} is a voltage independent constant for a given MOS structure, a $\phi_{MS} \neq 0$ simply gives rise to a parallel translation of the entire C–V characteristic along the voltage axis as illustrated in Fig. 4.2(b). Each capacitance value is totally determined by the U_S value inside the semiconductor and therefore occurs displaced ϕ_{MS} volts relative to the ideal case.

The actual ϕ_{MS} value for a given MOS structure is routinely computed from Eq. (4.2a) using the $\Phi_M' - \chi'$ appropriate for the system and the $(E_c - E_F)_\infty$ deduced from a knowledge of the doping concentration inside the semiconductor. The $\Phi_M' - \chi'$ values for a number of metal–silicon combinations have been determined experimentally and are listed in Table 4.1. The room temperature ϕ_{MS} for the commercially important Al–SiO_2–Si system is graphed as a function of doping in Fig. 4.3. Note from Fig. 4.3

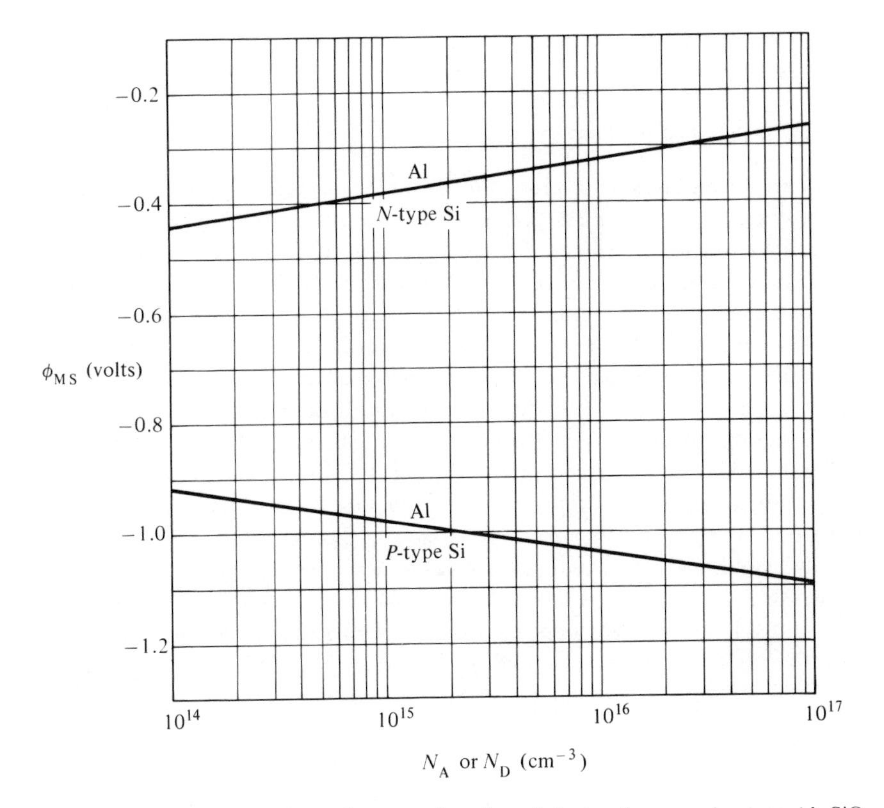

Fig. 4.3 Room temperature ϕ_{MS} values as a function of doping for n- and p-type Al–SiO_2–Si structures.

Table 4.1 Barrier Height Differences in Selected Metal–SiO_2–Si Structures.

Metal gate material	$\Phi_M - \chi = \Phi'_M - \chi'$ (eV)
Ag	0.73
Al	−0.11
Au	0.82
Cr	−0.06
Cu	0.63
Mg	−1.05
Sn	−0.83

(The above data is taken from S. Kar, *Solid-State Electronics* 18 (1975):169.)

and the $\Phi'_M - \chi'$ values listed in Table 4.1 that ϕ_{MS} is more often than not a negative quantity, especially for p-type devices, and is typically quite small—on the order of one volt or less.

4.2 MOBILE IONS IN THE OXIDE

The most perplexing and serious problem encountered in the development of MOS devices can be described as follows: First, the as-fabricated early (c. 1960) devices exhibited C–V characteristics which were sometimes shifted negatively by *tens* of volts with respect to the theoretical characteristics. Secondly, when subjected to bias-temperature (BT) stressing, a common reliability-testing procedure where a device is heated under bias to accelerate device-degrading processes, the MOS structures displayed a severe instability. The negative shift in the characteristics was increased additional tens of volts after the device was biased positively and heated up to 150°C or so. Negative bias-temperature stressing had the reverse effect; the C–V curve measured at room temperature after stressing shifted positively or toward the theoretical curve. In extreme cases the instability could even be observed by simply biasing the device at room temperature. One might sweep the C–V characteristics for a given device, go out to lunch leaving the device positively biased, and return to repeat the C–V measurement only to find the characteristics had shifted a volt or so toward negative biases. Note that the characteristics were always shifted in the direction opposite to the applied gate polarity and that the observed curves were always to the negative side of the theoretical curves. The nature and extent of the problem is nicely summarized in Fig. 4.4.

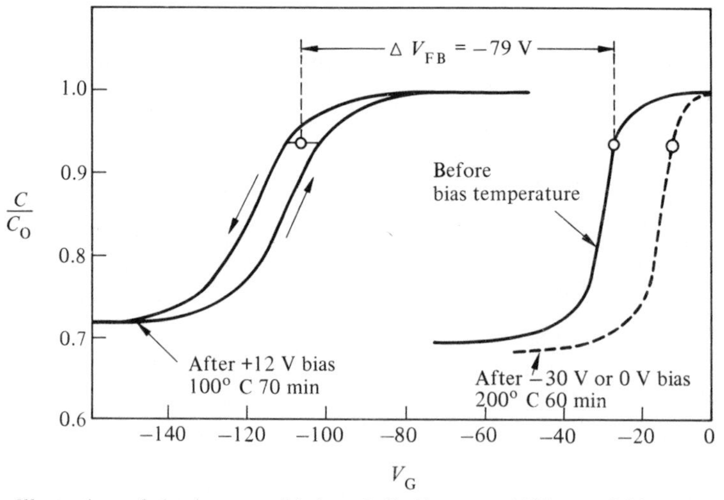

Fig. 4.4 Illustration of the large as-fabricated C–V curve shifting and bias-temperature instability observed with early MOS devices. All C–V curves were taken at room temperature; $x_o = 0.68\mu$. The arrows adjacent to the after + BT curves indicate the direction of the voltage sweep. (From D. R. Kerr et al., *IBM J. Res. & Dev.* 8: 376, September 1964; ©1964 by International Business Machines Corporation. Reprinted with permission.)

From a practical standpoint, the nonideality causing the as-fabricated translation and instability of the MOS device characteristics had to be identified and eliminated. A device whose effective operating point uncontrollably changes as a function of time is fairly useless. It is now well established that the large as-fabricated shifting and the related instability can be traced to mobile ions inside the oxide, principally Na^+. In the remainder of this section we will examine some of the considerations which led to this conclusion and will discuss fabrication procedures which have been subsequently instituted to minimize the problem.

To determine how charge centers in the oxide modify the characteristics of an MOS device, let us postulate the existence of a volume ionic charge distribution, $\rho_{ox}(x)$, which varies in an arbitrary manner (see Fig. 4.5(a)) across the width of the oxide layer. (Note from Fig. 4.5(a) that, for convenience in this particular analysis, *the origin of the x-coordinate has been relocated at the Metal–Oxide interface*.) With the addition of the charge centers, a portion of the gate voltage–U_S derivation presented in Section 2.4 is no longer valid and must be revised. Specifically, in lieu of Eqs. (2.33) to (2.35), one has, respectively,

$$\frac{d\mathscr{E}_{ox}}{dx} = \frac{\rho_{ox}(x)}{K_O\varepsilon_0} \tag{4.5}$$

$$\mathscr{E}_{ox}(x) = -\frac{dV_{ox}}{dx} = \mathscr{E}_{ox}(x_o) - \frac{1}{K_O\varepsilon_0}\int_x^{x_o} \rho_{ox}(x')\,dx' \tag{4.6}$$

and

$$\Delta V_{ox} = x_o\mathscr{E}_{ox}(x_o) - \frac{1}{K_O\varepsilon_0}\int_0^{x_o}\int_x^{x_o} \rho_{ox}(x')\,dx'\,dx \tag{4.7}$$

Now, since we are still excluding a plane of charge at the oxide–semiconductor interface, $\mathscr{E}_{ox}(x_o) = K_S\mathscr{E}_S/K_O$. Moreover, the double integral in Eq. (4.7) can be reduced to a single integral employing integration by parts. The indicated manipulations yield

$$\Delta V_{ox} = x_o'\mathscr{E}_S - \frac{1}{K_O\varepsilon_0}\int_0^{x_o} x\rho_{ox}(x)\,dx \tag{4.8}$$

Thus, for a structure which has ions in the oxide but is otherwise ideal,

$$V_G = \frac{kT}{q}U_S + x_o'\mathscr{E}_S - \frac{1}{K_O\varepsilon_0}\int_0^{x_o} x\rho_{ox}(x)\,dx \tag{4.9}$$

and

$$\boxed{\Delta V_G = (V_G - V_G')\big|_{\text{same }U_S} = -\frac{1}{K_O\varepsilon_0}\int_0^{x_o} x\rho_{ox}(x)\,dx} \tag{4.10}$$

The C–V curve voltage translation specified by Eq. (4.10) is of course in addition to the previously determined voltage translation due to a nonzero ϕ_{MS}.

An examination of Eq. (4.10) leads to some very interesting conclusions. For one, positive ions in the oxide would give rise to a negative shift in the C–V characteristics

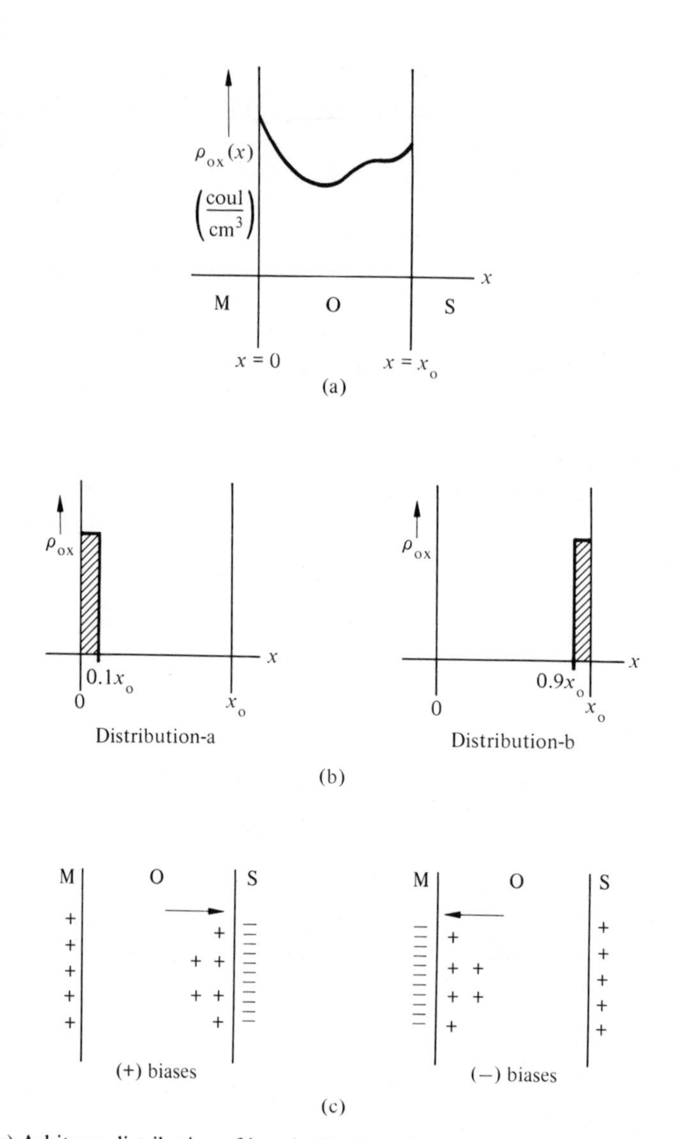

Fig. 4.5 (a) Arbitrary distribution of ions inside the oxide. (b) Two hypothetical charge distributions involving the same total number of ions situated near the metal (distribution–a) and near the semiconductor (distribution–b). (c) Expected motion of positive mobile ions within the oxide under (+) and (−) bias-temperature stressing.

as observed experimentally, while negative ions would give rise to a positive shift in disagreement with experimental observations. Furthermore, because the integrand in Eq. (4.10) varies as $x\rho_{ox}(x)$, ΔV_G is sensitive to the exact position of the ions in the oxide. If, for example, the same ionic charge per unit gate area, Q_M, is positioned (*a*) near the metal and (b) near the semiconductor as shown in Fig. 4.5(b), then one readily computes $\Delta V_G(a) = -0.05\ Q_M/C_o$ and $\Delta V_G(b) = -0.95\ Q_M/C_o$, where $C_o = K_O\varepsilon_0/x_o$. For the cited example the shift is predicted to be some 19 times larger when the ions are located near the oxide–semiconductor interface! Indeed, on the basis of the foregoing considerations, it is reasonable to speculate that a large as-fabricated negative shift in the measured C–V characteristics and the attendant instability is caused by positive ions in the oxide which move around or redistribute under bias-temperature stressing. The required ion movement, away from the metal for +BT stressing and toward the metal for −BT stressing, is in fact consistent with the direction of ion motion grossly expected from the repulsive/attractive action of other charges within the structure (see Fig. 4.5(c)).

Actual verification of the mobile ion model and identification of the culprit (the ionic specie) rivals some of the best courtroom dramas. The suspects were first indicted because of their past history and their accessibility to the scene of the crime. Long before the fabrication of the first MOS device, as far back as 1888, researchers had demonstrated that Na^+, Li^+, and K^+ ions could move through quartz, crystalline SiO_2, at temperatures below 250°C. Furthermore, alkali ions, especially sodium ions, were abundant in chemical reagents, in glass apparatus, on the hands of laboratory personnel, and in the tungsten evaporation boats used in forming the metallic gate. With the suspect identified, great care was taken to avoid alkali ion contamination in the formation of the MOS structure. The net result was devices which showed essentially no change in their C–V characteristics after they were subjected to either positive or negative biases for many hours at temperatures up to 200°C. Next, other carefully processed devices were purposely contaminated by rinsing the oxidized Si wafers in a dilute solution of NaCl (or LiCl) prior to metallization. As expected, the purposely contaminated devices exhibited severe instabilities under bias-temperature stressing. Finally, sodium was positively identified in the oxides of normally fabricated devices (no intentional contamination) through the use of the neutron activation technique; i. e., the oxides were bombarded with a sufficient number of neutrons to create a radioactive species of sodium. The analysis of the resultant radioactivity directly confirmed the presence of sodium within the oxide.

Although care to eliminate alkali ion contamination throughout the fabrication process does lead to stable MOS devices, most manufacturers encountered difficulties in attaining and maintaining the required degree of quality control in production-line facilities. For this reason modern-day processing typically includes a stabilization procedure which minimizes the effects of alkali ion contamination. Two different procedures have achieved widespread usage; namely, phosphorus stabilization and chlorine neutralization.

In phosphorus stabilization the oxidized Si wafer is simply placed in a phosphorus diffusion furnace for a short period of time. During the diffusion, as illustrated in Fig. 4.6(a), phosphorus enters the outer portion of the SiO_2 film and becomes incorporated into the bonding structure, thereby forming a new thin layer referred to as a phosphosilicate glass. At the diffusion temperature the sodium ions are extremely mobile and invariably wander into the phosphorus-laden region of the oxide. Once in the phosphosilicate glass the ions become trapped and stay trapped when the system is cooled to room temperature. In this way the alkali ions are "gettered" or drawn out of the major portion of the oxide, are positioned near the outer interface where they give rise to the

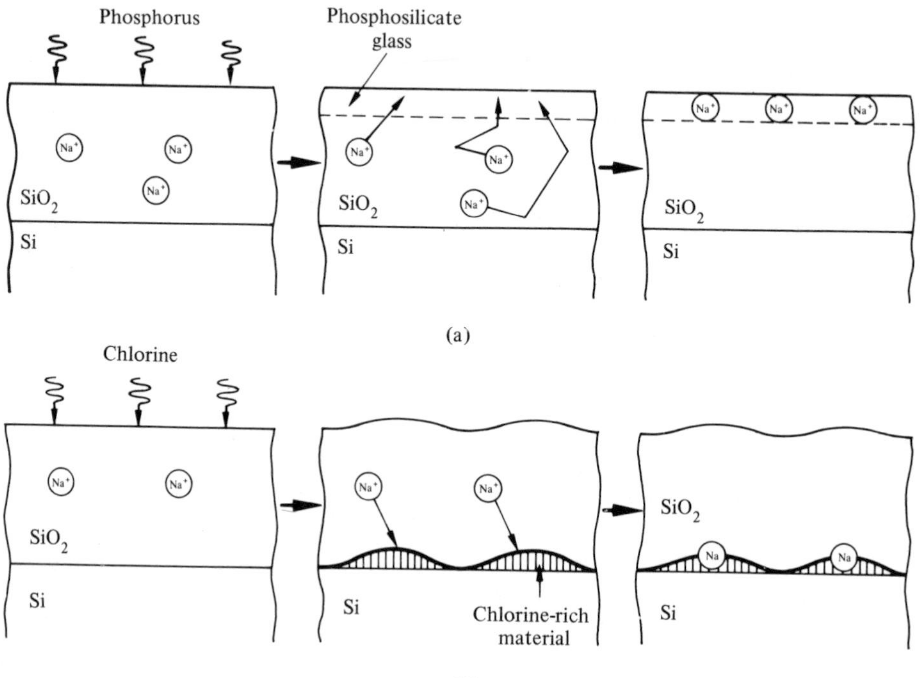

Fig. 4.6 Pictorial description of MOS stabilization procedures: (a) phosphorus stabilization; (b) chlorine neutralization.

least amount of as-fabricated $C-V$ curve shifting, and are held firmly in place during normal operating conditions. The phosphosilicate glass layer, it should be noted, also blocks any subsequent contamination associated with the gate metallization or other poststabilization processing steps.

Chlorine neutralization is a more recent stabilization procedure in which a small amount of chlorine in the form of HCl, Cl_2, trichloroethylene, or trichloroethane is introduced into the furnace ambient during the growth of the SiO_2 layer. As pictured in Fig. 4.6(b), the chlorine enters the oxide and reacts to form a new material, believed to be a chlorosiloxane, located at the oxide–silicon interface. The new material occupies pancake-shaped regions approximately 1 μ across and several hundred angstroms thick. Stabilization occurs when the ionic sodium migrates into the vicinity of the oxide–silicon interface, becomes trapped, and is *neutralized* by the chlorine in the chlorine-rich material near the interface. Neutral sodium has no effect, of course, on the MOS device characteristics. The chlorine neutralization procedure has become extremely popular because it does not require an additional processing step, is actually beneficial to the structure in other ways (increases the minority carrier lifetime and reduces stacking faults in the silicon), and is even useful in cleaning out furnace tubes.

4.3 THE FIXED OXIDE CHARGE

The gross perturbation associated with mobile ions in the oxide tended to obscure or cover up the effects of other deviations from the ideal. Indeed, with the successful elimination of the mobile ion problem, it became possible to perform a more exacting examination of the device characteristics. The results were rather intriguing. Even in structures free of mobile ions, and after correcting for $\phi_{MS} \neq 0$, the observed C–V characteristics *were still translated up to a few volts toward negative biases* relative to the theoretical characteristics. The possibility of mobile ion contamination had been eliminated because the structures were stable under bias-temperature stressing. Moreover, for a given set of fabrication conditions the observed ΔV_G was *completely reproducible*. Confirming data was obtained from devices independently fabricated by a number of workers at different locations. Subsequent testing (by etching the oxide away in small steps and through photomeasurements) revealed the unexplained ΔV_G shift was caused by a charge residing within the oxide very close to the oxide–semiconductor interface. Because this quasi-interfacial charge was reproducibly fabricated into the structure and was fixed in position under bias-temperature stressing, the nonideality became known as the "built-in" or "fixed" oxide charge.

In analyzing the effect of the fixed charge on the C–V characteristics, it is assumed the charge is located right at the oxide–semiconductor interface. If Q_F represents the fixed oxide charge per unit area at the oxide–semiconductor interface, then $Q_F = Q_{O\text{-}S}$ in Eq. (2.36) of Section 2.4 and Eq. (2.37) of the same section must be modified to read

$$D_{ox} = D_{semi|x=0} - Q_F \tag{4.11}$$

giving

$$V_G = \frac{kT}{q} U_S + x_o' \mathscr{E}_S - Q_F/C_o \tag{4.12}$$

where

$$C_o \equiv \frac{K_O \varepsilon_0}{x_o} = \text{oxide capacitance per unit area } (C_O = C_o A_G) \tag{4.13}$$

We therefore conclude the C–V translation due to the fixed oxide charge is

$$\boxed{\Delta V_G = -\frac{Q_F}{C_o}} \tag{4.14}$$

Note that, since Q_F is just a special case of the oxide charge distribution considered in the previous section, the above result can also be obtained from Eq. (4.10) by simply setting $\rho_{ox}(x) = Q_F \delta(x_o)$, where $\delta(x_o)$ is a delta-function positioned at the oxide–semiconductor interface.

From Eq. (4.14) it is obvious that, like the mobile ion charge, the fixed oxide charge must be *positive* to account for the negative ΔV_G's observed experimentally. Other relevant information about the fixed oxide charge can be summarized as follows.

1. The fixed charge is independent of the oxide thickness, the semiconductor doping concentration, and the semiconductor doping type (n or p).
2. The fixed charge varies as a function of the Si surface orientation; Q_F is largest on {111} surfaces, smallest on {100} surfaces, and the ratio of the fixed charge on the two surfaces is approximately 3 : 1.
3. Q_F is a strong function of the oxidation conditions such as the oxidizing ambient and furnace temperature. As displayed in Fig. 4.7, the fixed charge decreases more or less linearly with increasing oxidation temperatures. It should be emphasized, however, that only the *terminal* oxidation conditions are important. If, for example, a Si wafer is first oxidized in water vapor at 1000°C for 1 h, and then exposed to a dry O_2 ambient at 1200°C for a sufficiently long time to achieve a steady state condition (~5 min), the Q_F value will reflect only the dry oxidation process at 1200°C.
4. Annealing (that is, heating) of an oxidized Si wafer in an Ar or N_2 atmosphere for a time sufficient to achieve a steady state condition reduces Q_F to the value observed for dry oxidations at 1200°C. In other words, regardless of the oxidation conditions, the fixed charge can always be reduced to a minimum by annealing in an inert atmosphere.

The preceding experimental facts all provide clues to the physical origin of the fixed oxide charge. For one, although doping impurities from the semiconductor diffuse into the oxide during the high temperature oxidation process, the fixed charge was noted to be independent of the semiconductor doping concentration and doping type. The existence of ionized doping impurities within the oxide can therefore be eliminated as a possible source of Q_F. Secondly, the combination of the interfacial positioning of the fixed charge, the Si-surface orientation dependence, and the sensitivity of Q_F to the terminal oxidation conditions suggests that the fixed charge is intimately related to the oxidizing reaction at the Si–SiO_2 interface. In this regard, it should be understood that during the thermal formation of the SiO_2 layer, the oxidizing specie diffuses through the oxide and reacts at the Si–SiO_2 interface to form more SiO_2. Thus, the last oxide formed, the portion of the oxide controlled by the terminal oxidation conditions, lies closest to the Si–SiO_2 interface and contains the fixed oxide charge. From considerations such as these it has been postulated that the fixed oxide charge is due to *excess ionic silicon* which has broken away from the silicon proper and is waiting to react in the vicinity of the Si–SiO_2 interface when the oxidation process is abruptly terminated. Excess Si has in fact been detected in the oxide near the Si surface. Annealing in an inert atmosphere, a standard procedure for minimizing the fixed oxide charge, apparently reduces the excess reaction components and thereby lowers Q_F.

4.4 INTERFACIAL TRAPS

Judged in terms of their wide-ranging and degrading effect on the operational behavior of MIS devices, insulator–semiconductor interfacial traps must be considered the most important nonideality encountered in MIS structures. A common manifestation of a

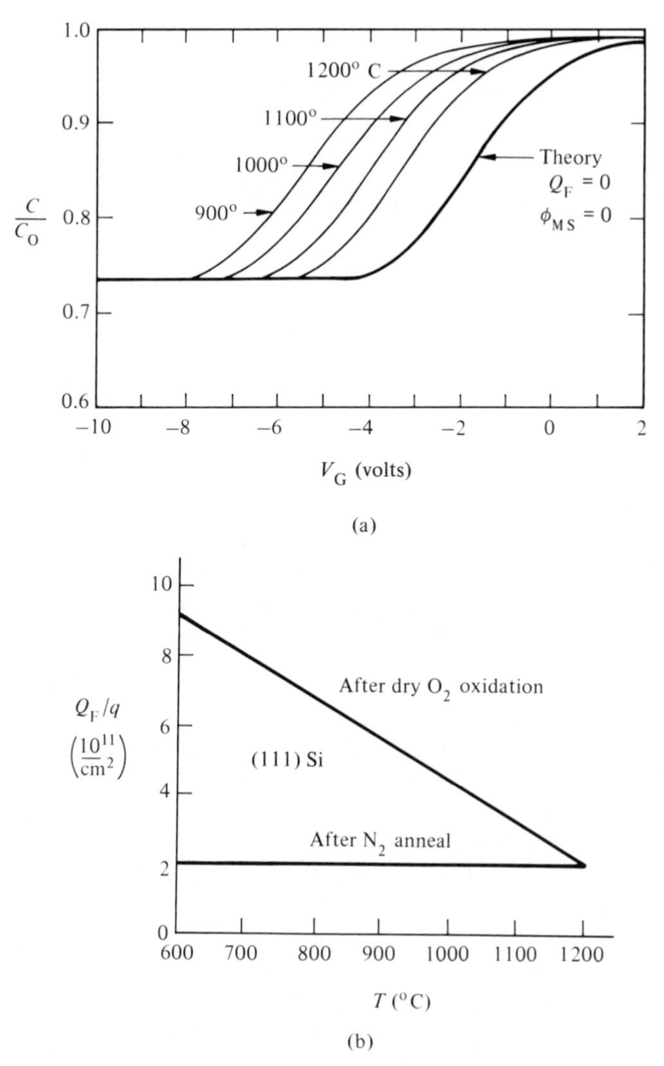

Fig. 4.7 Effect of the oxidation temperature and annealing on the fixed charge in MOS structures. (a) Measured C–V characteristics after dry O_2 oxidations at various temperatures ($x_o = 0.2\mu$, $N_D = 1.4 \times 10^{16}/\text{cm}^3$, (111) Si surface orientation). (b) Fixed charge concentrations—the so-called *oxidation triangle* specifying the expected Q_F/q after dry O_2 oxidation and after inert ambient annealing. (From B. E. Deal et al., *J. Electrochem. Soc.*, **114:** 266, March 1967. Reprinted by permission of the publisher, The Electrochemical Society, Inc.)

nonnegligible interfacial trap concentration within an MOS-C is the distorted or spread out nature of the C–V characteristics. This is nicely illustrated in Fig. 4.8, which displays two C–V curves derived from the same device before and after minimizing the number of Si–SiO_2 interfacial traps inside the structure.

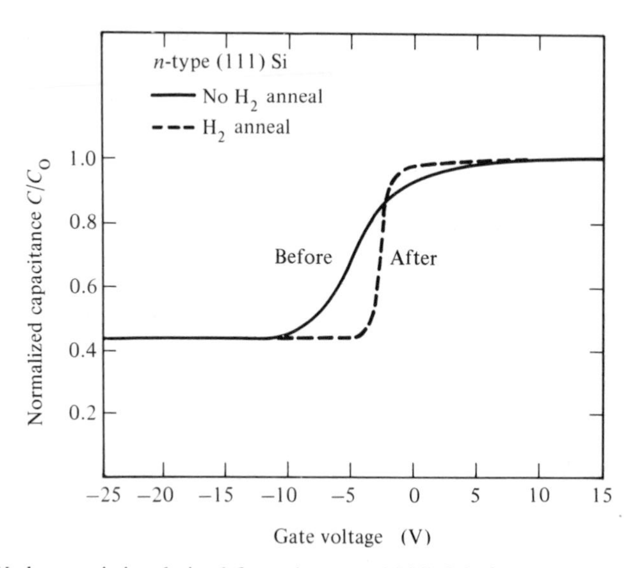

Fig. 4.8 *C–V* characteristics derived from the same MOS-C before (———) and after (– – –) minimizing the number of Si–SiO_2 interfacial traps inside the structure. (From R. R. Razouk and B. E. Deal, *J. Electrochem. Soc.*, **126:** 1573, 1979. Reprinted by permission of the publisher, The Electrochemical Society, Inc.)

Electrically, interfacial traps (also referred to as surface states or interface states) are allowed energy states in which electrons are localized in the vicinity of a material's surface. As modeled in Fig. 4.9(a), the interface states can and normally do introduce energy levels distributed throughout the forbidden band gap. Interface levels can also occur at energies greater than E_c or less than E_v, but such levels are usually obscured by the much larger density of conduction or valence band states.

Figure 4.9(b) to (d) provides some insight into the behavior and significance of the levels. When an *n*-bulk MOS-C is biased into inversion as shown in Fig. 4.9(b), the Fermi level at the surface lies close to E_v. For the given situation essentially all of the interfacial traps will be empty because, to a first-order approximation, all energy levels above E_F are empty and all energy levels below E_F are filled. Moreover, if the states are assumed to be donor-like in nature (that is, positively charged when empty and neutral when filled with an electron), the net charge per unit area associated with the interfacial traps, Q_{IT}, will be positive. Changing the gate bias to achieve depletion conditions (Fig. 4.9c) positions the Fermi level somewhere near the middle of the band gap at the surface. Since the interface levels always remain fixed in energy relative to E_c and E_v at the surface, depletion biasing obviously draws electrons into the lower interface state levels and Q_{IT} reflects the added negative charge: Q_{IT} (depletion) $<$ Q_{IT} (inversion). Finally, with the MOS-C accumulation biased (Fig. 4.9d), electrons fill most of the interfacial traps and Q_{IT} approaches its minimum value. The point here is that the interfacial traps charge and discharge as a function of bias, thereby affecting the charge distribution inside the device,

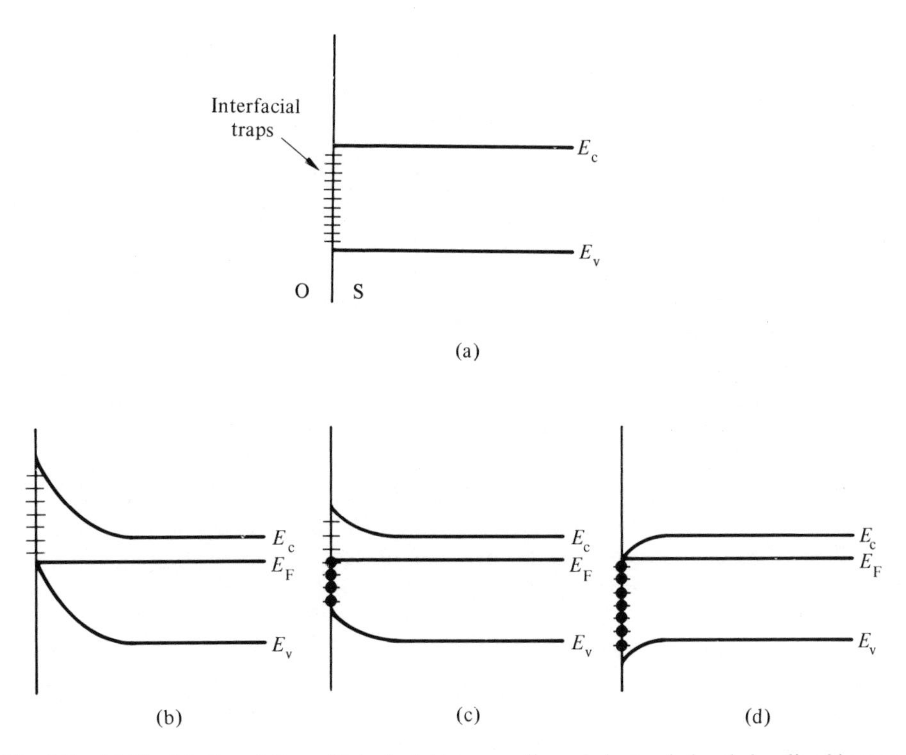

Fig. 4.9 (a) Electrical modeling of interfacial traps as allowed electronic levels localized in space at the oxide–semiconductor interface. (b)–(d) Filling of the interface levels under (b) inversion, (c) depletion, and (d) accumulation biasing in an n-type device.

the V_G–U_S relationship, and the device characteristics in an understandable but somewhat complex manner.

The general modification of the V_G–U_S relationship to account for interfacial traps is actually quite easy to establish. Since Q_{IT}, like Q_F, is located at the Si–SiO_2 interface, we can write, by direct analogy with the fixed charge analysis,

$$\boxed{\Delta V_G = -\frac{Q_{IT}(U_S)}{C_o}} \quad \begin{array}{l} C\text{–}V \text{ curve translation} \\ \text{due to interfacial traps} \end{array} \tag{4.15}$$

Combined with the considerations of the preceding paragraph, Eq. (4.15) helps to explain the form of the C–V characteristics presented in Fig. 4.8. Under inversion conditions, and for donor-like interfacial traps, Q_{IT} takes on its largest positive value and gives rise to a moderately large negative shift in the C–V characteristics. In progressing through depletion toward accumulation, Q_{IT} decreases, and the translation in the C–V curve likewise decreases as observed experimentally. Once in accumulation ΔV_G should continue to decrease and still remain negative according to the Fig. 4.9 model. The Fig. 4.8

data, on the other hand, exhibits an increasingly *positive* shift in the characteristics with increased accumulation biasing. This discrepancy can be traced to the donor-like assumption. In actual MOS devices the interfacial traps in the upper half of the band gap are believed to be acceptor-like in nature (that is, neutral when empty and negative when filled with an electron). Thus, upon reaching flat band (or roughly flat band), Q_{IT} passes through zero and becomes increasingly negative as more and more upper band gap states are filled with electrons. Qualitatively, then, we can explain the observed characteristics. A quantitative description would of course require both a detailed knowledge of the interfacial trap concentration versus energy and additional theoretical considerations to establish an explicit expression for Q_{IT} as a function of U_S.

Although models that detail the electrical behavior of the interfacial traps exist, the *physical origin* of the traps has not been totally clarified. The weight of experimental evidence, however, supports the view that the interfacial traps primarily arise from unsatisfied chemical bonds or so-called "dangling bonds" at the surface of the semiconductor. When the silicon lattice is abruptly terminated along a given plane to form a surface, one of the four surface-atom bonds is left dangling as pictured in Fig. 4.10(a). Logically, the thermal formation of the SiO_2 layer ties up some but not all of the Si-surface bonds. It is the remaining dangling bonds which become the interfacial traps (see Fig. 4.10(b)).

To add support to the foregoing physical model, let us perform a simple feasibility calculation. On a (100) surface there are 6.8×10^{14} Si atoms per cm^2. If 1/1000 of these form interfacial traps and one electronic charge is associated with each trap, the structure would contain a $Q_{IT}/q = 6.8 \times 10^{11}/\text{cm}^2$. Choosing an $x_o = 0.2\ \mu$ and substituting into Eq. (4.15) we obtain a ΔV_G (interfacial traps) $= 6.3$ V. Clearly, only a relatively small number of residual dangling bonds can significantly perturb the device characteristics and readily account for observed interfacial trap concentrations.

The overall interfacial trap concentration and the precise density of states as a function of energy (D_{IT}—units of states/cm^2-eV) are extremely sensitive to even minor fabrication details and vary significantly from device to device. Nevertheless, reproducible general trends have been recorded. The interfacial trap density, like the fixed oxide charge, is greatest on {111} Si surfaces, smallest on {100} surfaces, and the ratio of midgap states on the two surfaces is approximately 3 : 1. After oxidation in a dry O_2 ambient, D_{IT} is relatively high, $\sim 10^{11}$ to 10^{12} states/cm^2-eV at midgap, with the density decreasing for increased oxidation temperatures in a manner also paralleling the fixed oxide charge. Annealing at high temperatures ($\gtrsim 600°C$) in an inert ambient, however, does *not* minimize D_{IT}. Rather, as will be described shortly, annealing in the presence of hydrogen at relatively low temperatures ($\lesssim 500°C$) minimizes D_{IT}. D_{IT} at midgap after an ideal interface state anneal is roughly $\sim 10^{10}/\text{cm}^2$-eV and the distribution of states as a function of energy is of the form sketched in Fig. 4.11. As shown in this figure, the interfacial trap density is more or less constant over the midgap region and increases rapidly as one approaches the band edges. Lastly, the states near the two band edges are usually about equal in number and opposite in their charging character; that is, states near the conduction and valence bands are believed to be acceptor-like and donor-like in nature, respectively.

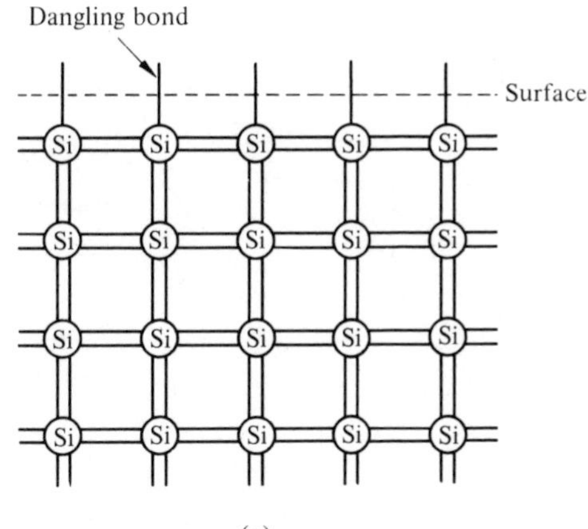

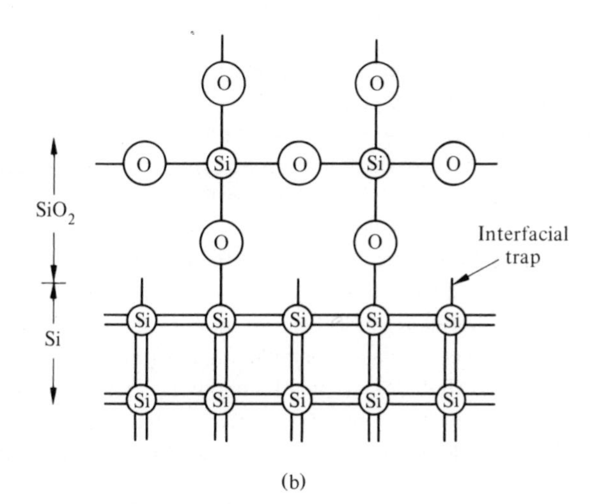

Fig. 4.10 Physical model for the interfacial traps. (a) "Dangling bonds" which occur when the Si lattice is abruptly terminated along a given plane to form a surface. (b) Postoxidation dangling bonds (relative number greatly exaggerated) that become the interfacial traps. (Part (b) adapted from B. E. Deal, *J. Electrochem. Soc.*, **121:** 198C (June 1974))

The very important annealing of MOS structures to minimize the interfacial trap concentration is routinely accomplished in one of two ways, namely, through postmetallization annealing or hydrogen (H_2) ambient annealing. In the postmetallization process, which requires a chemically active gate material such as Al or Cr, the metallized structure is simply placed in a nitrogen ambient at ~500°C for 5 to 10 min. During the formation of MOS structures, minute amounts of water vapor inevitably become adsorbed on the SiO_2 surface. At the postmetallization annealing temperature the active gate

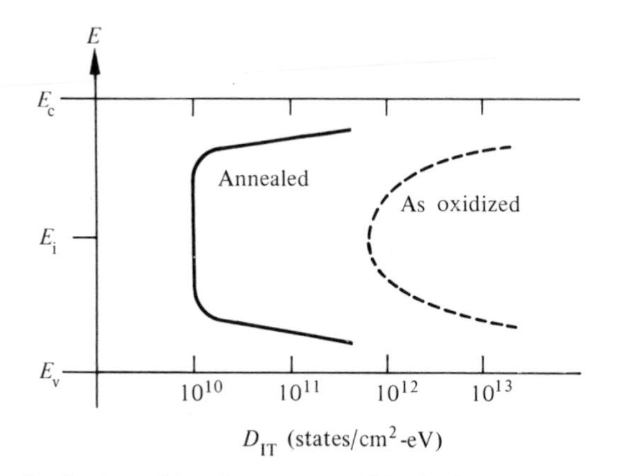

Fig. 4.11 Energy distribution of interface states within the band gap. General form and magnitude of the interfacial trap density observed before and after an interface state anneal.

material reacts with the water vapor on the oxide surface to release a hydrogen specie thought to be atomic hydrogen. As pictured in Fig. 4.12, the hydrogen specie subsequently migrates through the SiO_2 layer to the Si–SiO_2 interface where it attaches itself to a dangling Si bond, thereby making the bond electrically inactive. The hydrogen ambient process operates on a similar principle, except the hydrogen is supplied directly in the ambient and the structure need not be metallized.

Even though we originally stated that the interfacial trap problem was of paramount importance, it is very difficult to convey the true significance and scope of the problem. Bluntly stated, if the thermal oxide didn't tie up most of the dangling Si bonds, and if an annealing process were not available for reducing the remaining bonds or interfacial traps to an acceptable level, MOS devices would merely be a laboratory curiosity. Indeed, high interfacial trap concentrations have severely stunted the development of other insulator–semiconductor systems. Moreover, in this section we have only scratched the

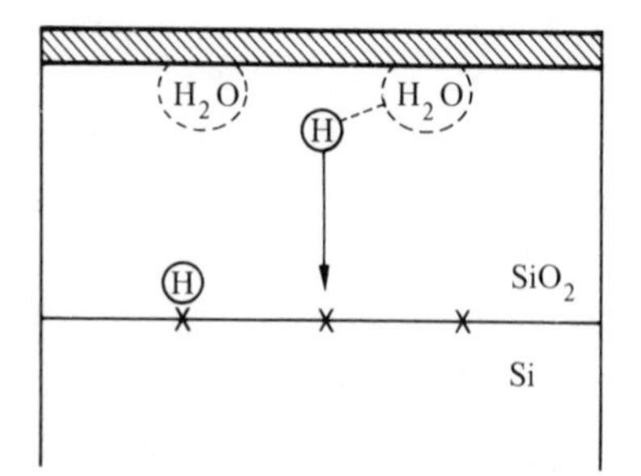

Fig. 4.12 Model for the annihilation of interface states during the postmetallization annealing process. (H)'s represent the active hydrogen species involved in the process; X's represent interface states.

surface so to speak. Interfacial traps degrade the operational behavior of MIS devices in a number of ways which can be understood only after increased knowledge is gained about the workings of MIS devices.

4.5 SUMMARY AND CONCLUDING COMMENTS

Stated concisely, real MOS devices are not intrinsically perfect. The devices are in fact intrinsically imperfect. In this chapter we cited and scrutinized four of the most commonly encountered deviations from the ideal; specifically, the metal–semiconductor work-function difference, mobile ions in the oxide, the fixed oxide charge, and interfacial traps. The topics covered are of major importance and quite adequately illustrate analytical procedures. It should be understood, however, that there are other nonidealities that can become very important under certain circumstances.

The effect of the analyzed nonidealities on the V_G–U_S relationship is summarized by Eq. (4.16).

$$\boxed{\Delta V_G = (V_G - V_G')\big|_{\text{same } U_S} = -\frac{Q_M\gamma_M}{C_o} - \frac{Q_F}{C_o} - \frac{Q_{IT}(U_S)}{C_o} + \phi_{MS}} \tag{4.16}$$

where

$$\gamma_M \equiv \int_0^{x_o} x\rho_{ox}(x)\,dx \Big/ x_o \int_0^{x_o} \rho_{ox}(x)\,dx \tag{4.17}$$

In writing Eq. (4.16) we recast the mobile ion contribution [Eq. (4.10)] to emphasize the similarity between the three terms associated with charges positioned in or adjacent to the oxide. γ_M is a unitless quantity representing the centroid of the mobile ion charge in the oxide normalized to the width of the oxide layer; $\gamma_M = 0$ if the mobile ions are all at the metal–oxide interface, while $\gamma_M = 1$ if the mobile ions are all piled up at the Si–SiO_2 interface. As a general rule, Q_M, Q_F, and ϕ_{MS} all lead to a negative parallel translation of the C–V characteristics along the voltage axis relative to the ideal theory. ΔV_G due to Q_{IT}, on the other hand, can be either positive or negative, depends on the applied bias, and tends to distort or spread out the characteristics.

Finally, a key point of the discussion was that procedures do exist for minimizing the net effect of nonidealities in MOS structures. Although constant checks must be run to maintain quality control, manufacturers today routinely make near-ideal MOS devices. It should be noted, however, that whereas the described or similar nonidealities apply to all MIS structures, the specific minimization procedures outlined herein apply only to the metal–SiO_2–Si system.

PROBLEMS

4.1 With the proper choice of gate material and the Si doping concentration it is possible to build an MOS-C with a $\phi_{MS} = 0$. Restricting the Si doping to be in the range

$$10^{14}/\text{cm}^3 \le N_A \quad \text{or} \quad N_D \le 10^{17}/\text{cm}^3,$$

and assuming $T = 23°C$ operation, identify the gate material-doping concentration combination(s) which gives rise to a $\phi_{MS} = 0$. Employ $\Phi'_M - \chi'$ values given in Table 4.1.

4.2 As previously mentioned, the gate material in many MOS devices is actually heavily doped polycrystalline Si. Consider a Si–gate MOS-C where $E_F = E_c$ in the heavily doped gate and $E_c - E_F = 0.2$ eV in the nondegenerately doped silicon substrate. Assuming the structure to be ideal (other than an obvious $\phi_{MS} \neq 0$),

(a) Sketch the energy band diagram for the Si–gate MOS-C under flat band conditions.

(b) What is the "metal"–semiconductor workfunction difference for the cited Si–gate MOS-C?

(c) Will the given MOS-C be accumulation, depletion, or inversion biased when $V_G = 0$? Explain.

(d) A nonzero fixed oxide charge typically appears at a $Si–SiO_2$ interface. What effect would a $Q_F \neq 0$ at the *Si–gate*–SiO_2 interface have upon the C–V_G characteristics derived from the MOS-C? Explain.

4.3 In analyzing the effect of the fixed charge (Q_F) it is commonly assumed the charge lies right at the oxide–semiconductor interface. Suppose the charge was actually distributed a short distance into the oxide from the $Si–SiO_2$ interface.

(a) Write down Eq. (4.14) which gives the ΔV_G in the C–V_G characteristics associated with a plane of charge located right at the $Si–SiO_2$ interface.

(b) Determine the expected ΔV_G shift caused by an equivalent amount of charge distributed in a linearly increasing fashion from zero at a distance Δx from the $Si–SiO_2$ interface to $2Q_F/\Delta x$ at the $Si–SiO_2$ interface.

(c) Compute $\Delta V_G(b)/\Delta V_G(a)$ using $\Delta x = 20$Å $= 2 \times 10^{-7}$ cm and $x_o = 0.1\ \mu = 10^{-5}$ cm.

4.4 In this problem we examine the shift in the MOS-C C–V_G characteristics resulting from certain specific nonidealities.

(a) An $N_D = 10^{15}/cm^3$ doped MOS structure is fabricated with a chromium gate. If $Q_M = 0$, $Q_F = 0$, and $Q_{IT} = 0$, determine the gate voltage required to achieve flat band. Use $kT = 0.0255$ eV, $n_i = 8.60 \times 10^9/cm^3$ and $E_G = 1.12$ eV.

(b) An MOS-C is found to possess a uniform distribution of sodium ions in the oxide; i.e., $\rho_{ox}(x) = \rho_0 =$ constant for all x in the oxide. Compute the ΔV_G shift resulting from this distribution of ions if $\rho_0/q =$ ion concentration $= 10^{18}/cm^3$, $x_o = 0.1\ \mu$ and $K_O = 3.9$.

(c) Thermal oxidation of a (111) oriented silicon substrate is followed by an N_2 anneal at 1000°C for a time sufficient to achieve a steady state condition. After depositing Al the structure is next postmetallization annealed. The completed MOS-C is found to be stable under bias-temperature stressing. Determine the expected flat band voltage if $T = 23°C$, $x_o = 0.1\ \mu$, and $N_A = 10^{14}/cm^3$. Be sure to consider all the nonidealities discussed in the chapter.

4.5 If interfacial traps are associated with residual "dangling bonds" at the Si surface, and assuming the number of residual "dangling bonds" is proportional to the number of Si surface atoms, which silicon surface plane, (100) or (110), would be expected to exhibit the higher density of interfacial traps? Record all work leading to your answer.

4.6 A rather unusual p-bulk MOS-C is found to have interfacial traps at only one band gap energy, E_{IT}, located right at midgap (see Fig. P4.6). Assuming a high frequency C–V_G measurement and an otherwise ideal MOS-C,

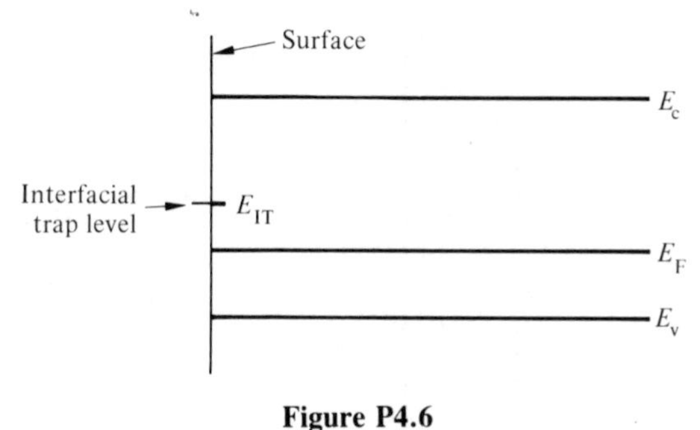

Figure P4.6

(a) Sketch the general form of the expected MOS-C C–V_G characteristics if the interface states giving rise to the E_{IT} level are donor-like in nature and perturb only the V_G–U_S relationship. (Also assume the number of states is sufficiently large to perturb the idealized characteristic.)

(b) Repeat (a) if the interface states are acceptor-like in nature.

(c) Repeat (a) assuming the interface states are donor-like but the energy level is located very close to the conduction band (say $E_c - E_{IT} = 0.001$ eV).

(d) Repeat (a) for donor-like interfacial traps which introduce an energy level very close to the valence band (say $E_{IT} - E_v = 0.001$ eV).

Include a few words of explanation if necessary to convey your thought process and to prevent a misinterpretation of your sketches.

4.7 One method of detecting interfacial traps, especially interfacial traps with energy levels moderately close to the band edges, is to examine the effect of reduced system temperatures on the MOS-C C–V_G characteristics.

(a) What happens to the positioning of the Fermi level in the band gap of an *n-type* semiconductor when the temperature is lowered below room temperature? (It may be helpful to review Problem 2.13 of Volume I.)

(b) Looking at the n-type semiconductor under flat band conditions and assuming an MOS structure with a nonnegligible number of interface states distributed throughout the band gap, what happens to the interface state charge (Q_{IT}) when the temperature is lowered below room temperature? Specifically, relative to room temperature, does Q_{IT} increase negatively, decrease, stay the same—or what? Record your reasoning. (*Note*: Your answer here should be the same whether the states are donor-like or acceptor-like in nature.)

(c) Based on the part (b) answer, and assuming all MOS parameters other than Q_{IT} are essentially unaffected by temperature changes, what will happen to the C–V_G characteristics as the system temperature is progressively reduced? Record your reasoning.

4.8 *MOS-C REVIEW PROBLEM*: A totally dimensioned energy band diagram for an MOS-C under a specific gate bias is shown in Fig. P4.8. The MOS-C has been carefully fabricated to minimize sodium ions and interfacial traps (i.e., $Q_M = 0$ and $Q_{IT} = 0$); ϕ_{MS} and Q_F are

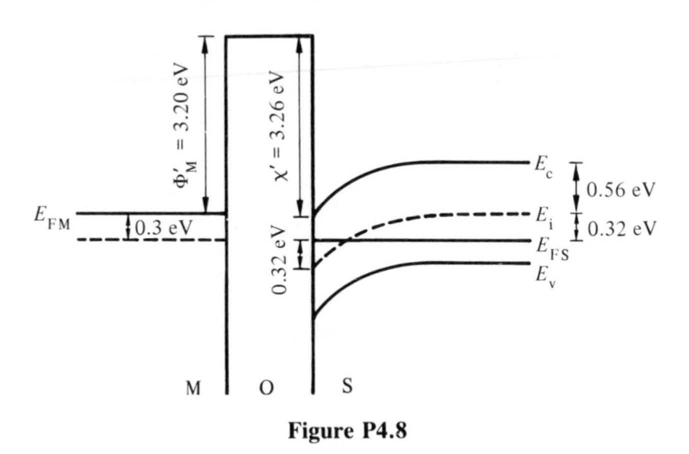

Figure P4.8

not necessarily zero. The device is maintained at T = 23°C and kT/q = 0.0255 V, n_i = 8.60 × $10^9/\text{cm}^3$, K_S = 11.8, K_O = 3.9 and x_o = 0.1 μ.

(a) Sketch the normalized electrostatic potential, U, inside the semiconductor as a function of position. (U = 0 in the semiconductor bulk.)

(b) Sketch the electric field, $\mathscr{E}$, inside the semiconductor as a function of position.

(c) Sketch the hole concentration, p, inside the semiconductor as a function of position.

(d) Determine N_A.

(e) Determine V_G. (Give the proper sign.)

(f) What is the total voltage drop (ΔV_{ox}) across the oxide?

(g) What is the total voltage drop across the semiconductor?

(h) Is the device accumulation, depletion, or inversion biased? Explain.

(i) Invoking the delta-depletion approximation, determine the normalized small signal capacitance, C/C_O, at the applied bias point.

(j) What is the metal–semiconductor workfunction difference (ϕ_{MS})?

(k) Is Q_F = 0 or is $Q_F \neq$ 0? Explain. (It is not necessary to actually determine Q_F to answer this question.)

5 / MOS Field Effect Transistors

Whereas the MOS-capacitor is very useful for diagnosing structural ills and analyzing the operational behavior of the MOS system, it is the MOS-transistor which occupies the position of prominence in the world of practical applications. There are literally hundreds of MOS-transistor circuits in production today, ranging from rather simple logic gates used in digital signal processing to custom designs with both logic and memory functions on the same silicon chip. The MOS-transistor is found in a mind-boggling number of electronic systems including the now commonplace hand-held calculator. Initially the MOS-transistor was identified by several competing acronyms; namely, MOST—Metal–Oxide–Semiconductor Transistor, IGFET—Insulated Gate Field Effect Transistor, and MOSFET–Metal–Oxide–Semiconductor Field Effect Transistor. (PIGFET and MISFET were even suggested with a smile at one time or another.) With the passage of time, however, the transistor structure has commonly come to be known as the MOSFET. In this chapter we examine the basic theory of MOSFET operation, beginning with a qualitative discussion of operational principles, progressing through a quantitative analysis of the dc characteristics and an inspection of related considerations, and concluding with an investigation of the ac response. The topic development, it should be noted, closely parallels the J-FET presentation contained in Chapter 1 of this volume. The reader may find it helpful, therefore, to spend a few moments performing a preliminary review of the J-FET presentation.

5.1 QUALITATIVE THEORY OF OPERATION

Figure 5.1 shows a cross-sectional view of the basic MOS-transistor structure. Physically, the MOSFET is essentially nothing more than an MOS-capacitor with two p-n junctions placed immediately adjacent to the region of the semiconductor controlled by the MOS-gate. The Si substrate can be either p-type (as pictured) or n-type; p^+ junction islands are of course required in n-bulk devices. Also shown in Fig. 5.1 are the standard terminal and dc voltage designations. As in the J-FET, carriers enter the structure through the source (S), leave through the drain (D), and are subject to the control or gating action of the gate

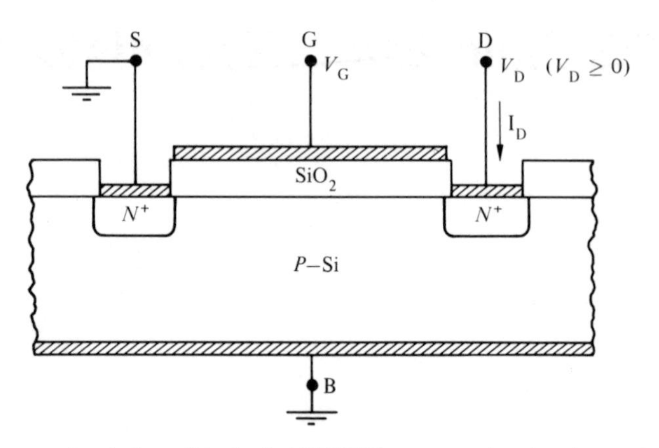

Fig. 5.1 Cross-sectional view of the basic MOSFET structure showing the terminal designations and standard biasing conditions.

(G). The voltage applied to the gate relative to ground is V_G, while the drain voltage relative to ground is V_D. Unless stated otherwise, we will assume herein that the source and back are grounded. Please note that under normal operational conditions the drain bias is always such as to reverse bias the drain p-n junction ($V_D \geq 0$ for the Fig. 5.1 device). Finally, consistent with the source and drain designations, the drain current (I_D) for the p-bulk device is taken to be positive when flowing into the drain terminal.

To establish the basic principles of MOSFET operation, let us first set $V_D = 0$ and examine the situation inside the structure as a function of the applied gate voltage. When V_G is accumulation or depletion biased ($V_G \leq V_T$, where V_T is the depletion–inversion transition-point voltage), the gated region between the source and drain islands contains either an excess or deficit of holes and very few electrons. Thus, looking along the surface between the n^+ islands under the cited conditions one effectively sees an open circuit. When V_G is inversion biased ($V_G > V_T$), however, an inversion layer containing mobile electrons is formed adjacent to the Si surface. Now looking along the surface between the n^+ islands one sees an induced "n-type" region (the inversion layer) or conducting *channel* connecting the source and drain islands, as pictured in Fig. 5.2(a). Naturally, the greater the inversion bias, the greater the pile-up of electrons at the Si surface and the greater the conductance of the inversion layer. An inverting gate bias, therefore, creates or induces a source-to-drain channel and determines the maximum conductance of the channel.

Turning next to the action of the drain bias, suppose an inversion bias $V_G > V_T$ is applied to the gate and the drain voltage is increased in small steps starting from $V_D = 0$. At $V_D = 0$ the situation inside the device is as previously pictured in Fig. 5.2(a), thermal equilibrium obviously prevails, and the drain current is identically zero. With V_D stepped to small positive voltages, the surface channel merely acts like a simple resistor and a drain current proportional to V_D begins to flow into the drain. The portion of the I_D–V_D relationship corresponding to small V_D biases is shown as the line from point O to point

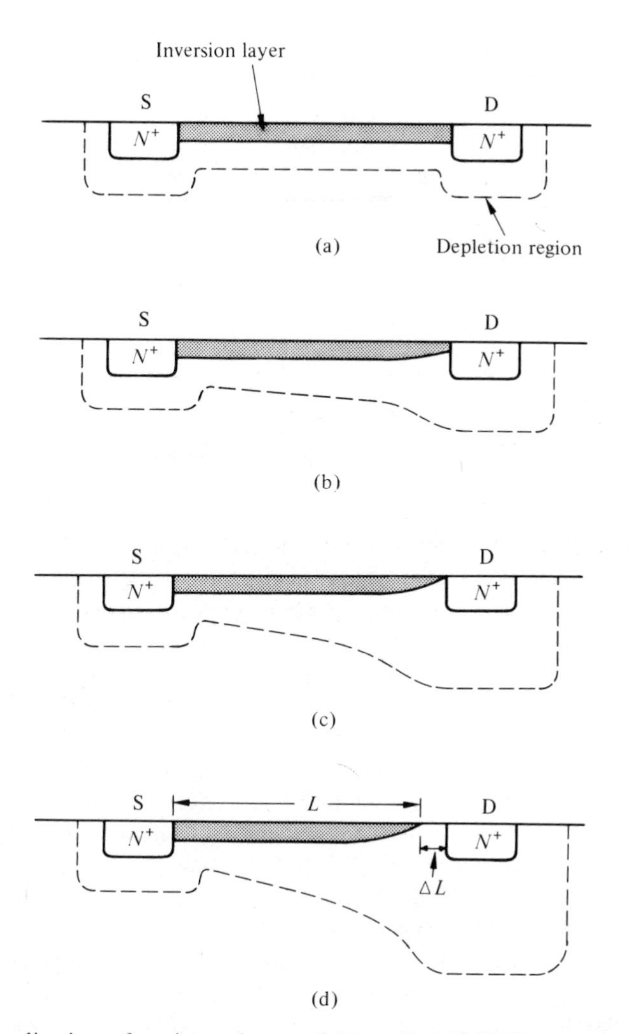

Fig. 5.2 Visualization of various phases of $V_G > V_T$ MOSFET operation. (a) $V_D = 0$; (b) channel (inversion layer) narrowing under moderate V_D biasing; (c) pinch-off; and (d) post-pinch-off ($V_D > V_{Dsat}$) operation. (Note that the inversion layer widths, depletion widths, etc. are not drawn to scale.)

A in Fig. 5.3. Any $V_D > 0$ bias, it should be interjected, simultaneously reverse biases the drain *p*-*n* junction, and the resulting reverse bias junction current flowing into the Si substrate does contribute to I_D. In well-made devices, however, the junction leakage current is totally negligible compared to the channel current, provided V_D is less than the junction breakdown voltage.

Once V_D is increased above a few tenths of a volt, the device enters a new phase of operation. Specifically, the voltage drop from the source to the drain associated with the

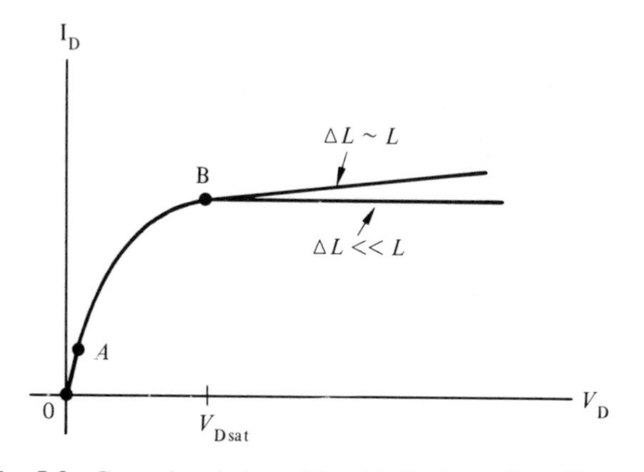

Fig. 5.3 General variation of I_D with V_D for a given $V_G > V_T$.

flow of current in the channel starts to negate the inverting effect of the gate. As pictured in Fig. 5.2(b), the depletion region widens in going down the channel from the source to the drain and the number of inversion layer carriers correspondingly decreases. The reduced number of carriers decreases the channel conductance, which in turn is reflected as a decrease in the slope of the observed I_D–V_D characteristic. Continuing to increase the drain voltage causes an ever-increasing depletion of the channel and the systematic slope-over in the I_D–V_D characteristic noted in Fig. 5.3. The greatest decrease in channel carriers occurs of course near the drain, and eventually the inversion layer completely vanishes in the near vicinity of the drain (see Fig. 5.2(c)). Like the analogous situation in the J-FET, the disappearance of the conducting channel adjacent to the drain in the MOSFET is referred to as "*pinch-off*." When the channel pinches off inside the device, the point B is reached on the Fig. 5.3 characteristic; that is, the slope of the I_D–V_D characteristic becomes approximately zero.

For drain voltages in excess of the pinch-off voltage, V_{Dsat}, the pinched-off portion of the channel widens from just a point into a depleted channel section ΔL in extent (see Fig. 5.2(d)). Since the pinched-off ΔL section absorbs most of the voltage drop in excess of V_{Dsat}, and given $\Delta L \ll L$, the source to pinch-off region of the device will be essentially identical in shape and will have the same endpoint voltages for all $V_D \geq V_{Dsat}$. When the shape of a conducting region and the potential applied across the region do not change, the current through the region must also remain invariant. Thus, I_D remains approximately constant for drain voltages in excess of V_{Dsat} provided $\Delta L \ll L$. If ΔL is comparable to L, the case in a large percentage of modern-day MOSFET's, the same voltage drop (V_{Dsat}) will appear across a shorter channel ($L - \Delta L$) and, as noted in Fig. 5.3, the post-pinch-off I_D in such devices will increase somewhat with increasing $V_D > V_{Dsat}$.

Up to this point we have examined the response of the MOSFET to the separate manipulation of the gate and drain biases. To complete the discussion, to establish a complete set of I_D–V_D characteristics, it is necessary to combine the results derived from the separate considerations. Clearly, for $V_G \leq V_T$, the gate bias does not create a surface

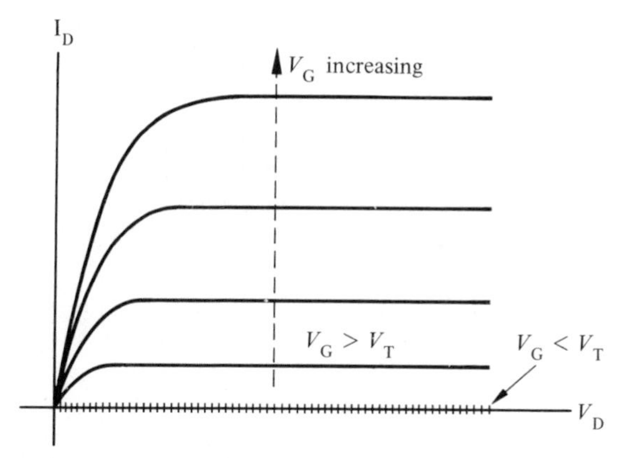

Fig. 5.4 General form of the I_D–V_D characteristics expected from a long channel ($\Delta L \ll L$) MOSFET.

channel and $I_D \simeq 0$ for all drain biases below the junction breakdown voltage. For all $V_G > V_T$ a characteristic of the form shown in Fig. 5.3 will be observed. Since the conductance of the channel increases with increasing V_G, it follows that the initial slope of the I_D–V_D characteristic will likewise increase with increasing V_G. Moreover, the greater the number of inversion layer carriers present when $V_D = 0$, the larger the drain voltage required to achieve pinch-off. Thus V_{Dsat} must increase with increasing V_G. From the foregoing arguments one concludes that the variation of I_D with V_D and V_G must be of the form displayed in Fig. 5.4.

To complete the discussion we should note that when the channel carriers are electrons the MOSFET is referred to as an *n-channel* device; when the channel carriers are holes the MOSFET is designated a *p-channel* device. Also, V_T, a parameter which obviously plays a prominant role in transistor operation, is commonly called the *threshold* or *turn-on* voltage in MOSFET analyses. The transistor begins to carry current—turns-on—at the onset of inversion.

5.2 QUANTITATIVE I_D–V_D RELATIONSHIPS

5.2.1 The Effective Mobility

In establishing quantitative expressions for the MOSFET dc characteristics it is necessary to first introduce a parameter known as the "effective mobility." The carrier mobilities, μ_n and μ_p, were first described in Section 3.1 of Volume I and were noted to be a measure of the ease of carrier motion within a semiconductor crystal. In the semiconductor bulk, that is, at a point far removed from the semiconductor surface, the carrier mobilities are typically determined by the amount of lattice scattering and ionized impurity scattering taking place inside the material. For a given temperature and semiconductor doping, these bulk mobilities (μ_n and μ_p) are well defined and well documented material constants.

Carrier motion in a MOSFET, however, takes place in a surface inversion layer where the gate-induced electric field acts so as to accelerate the carriers toward the surface. The inversion layer carriers therefore experience motion impeding collisions with the Si surface (see Fig. 5.5) in addition to lattice and ionized impurity scattering. The additional surface scattering mechanism lowers the mobility of the carriers, with the carriers constrained nearest the Si surface experiencing the greatest reduction in mobility. The resulting average mobility of the inversion layer carriers is called the *effective mobility* and is given the symbol $\overline{\mu}_n$ or $\overline{\mu}_p$.

Seeking to establish the formal mathematical expression for the effective mobility, let us consider an n-channel device with the structure and dimensions specified in Fig. 5.6. Let x be the depth into the semiconductor measured from the oxide–semiconductor interface, y the distance along the channel measured from the source, $x_c(y)$ the channel depth, $n(x, y)$ the electron concentration at a point (x, y) in the channel, and $\mu_n(x, y)$ the mobility of carriers at the (x, y) point in the channel. Invoking the standard averaging procedure, the effective mobility of carriers an arbitrary distance y from the source is simply given by

$$\overline{\mu}_n = \frac{\int_0^{x_c(y)} \mu_n(x, y) n(x, y)\, dx}{\int_0^{x_c(y)} n(x, y)\, dx} \tag{5.1}$$

Note also that the total electronic charge/cm^2 in the channel at a given y is

$$Q_N(y) = -q \int_0^{x_c(y)} n(x, y)\, dx \tag{5.2}$$

Consequently, one can alternatively write

$$\overline{\mu}_n = -\frac{q}{Q_N(y)} \int_0^{x_c(y)} \mu_n(x, y) n(x, y)\, dx \tag{5.3}$$

If the drain voltage is small, the channel depth and carrier charge will be more or less uniform from source to drain and the effective mobility will be essentially the same for all y values. When the drain voltage becomes large, on the other hand, x_c and Q_N vary with position, and it is reasonable to expect that $\overline{\mu}_n$ likewise varies somewhat in going down the channel from the source to the drain. To obtain tractable closed-form results for the I_D–V_D characteristics, however, it is necessary to neglect the cited y-dependence.

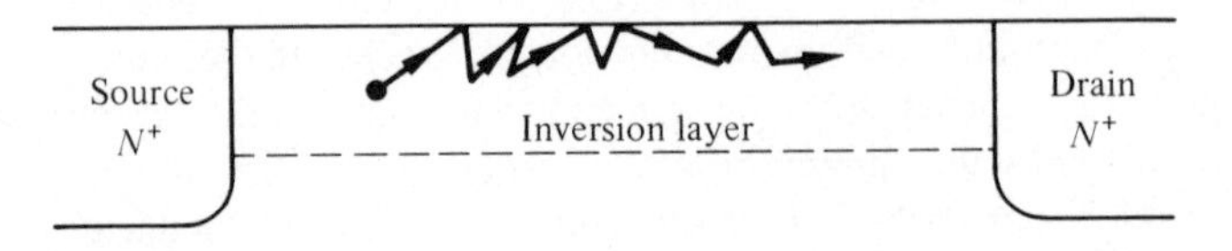

Fig. 5.5 Visualization of surface scattering at the Si–SiO_2 interface.

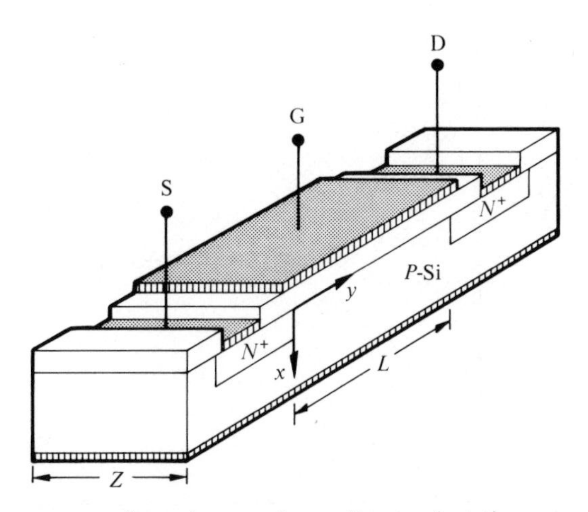

Fig. 5.6 Device structure, dimensions, and coordinate orientations assumed in the quantitative analysis.

Thus, *herein we will subsequently consider* $\bar{\mu}_n$ *to be a device parameter which is independent of y and the applied drain voltage* V_D.

Relative to the dependence of $\bar{\mu}_n$ on the applied *gate* voltage, increased inversion biasing places more carriers closer to the oxide–semiconductor interface and increases the electric field acting on the carriers. This combination of effects enhances surface scattering and thereby lowers the average carrier mobility; $\bar{\mu}_n$ therefore decreases with increased inversion biasing. The exact $\bar{\mu}_n$ versus V_G dependence varies from device to device but generally follows the form displayed in Fig. 5.7. Also note from Fig. 5.7 that the surface scattering phenomenon can be rather significant, giving rise to effective mobilities considerably below the bulk μ_n.

5.2.2 General Analysis

As in the case of the MOS-C capacitance–voltage characteristics, there exists a hierarchy of MOSFET I_D–V_D theories providing increased accuracy at the expense of increased complexity. We will examine two of the theories; namely, the "square-law" theory and the "bulk-charge" theory. The former provides very simple relationships; the latter is a much more accurate representation of reality. Interestingly, all but the final derivational steps in the two theories are identical. The common derivational steps are presented in this subsection.

For gate voltages above turn-on, $V_G > V_T$, and drain voltages below pinch-off, $0 \le V_D \le V_{Dsat}$, the derivation of the desired I_D–V_D relationship proceeds as follows: In general one can write

$$\mathbf{J}_N = q\mu_n n\mathscr{E} + qD_N\nabla n \tag{5.4}$$

Within the conducting channel the current is flowing almost exclusively in the y-direction.

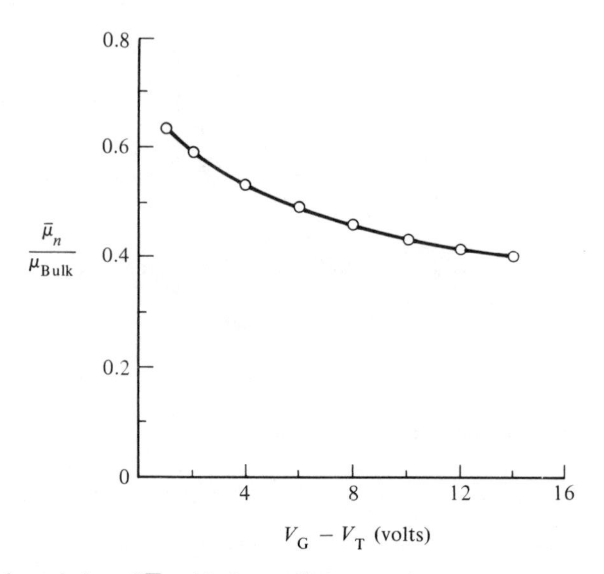

Fig. 5.7 Sample variation of $\bar{\mu}_n$ with the applied gate voltage ($V_D \simeq 0$). (Data from S. C. Sun and J. D. Plummer, *IEEE Trans. on E. D.*: **ED-27:** 1497, August, 1980).

Moreover, paralleling the J-FET analysis, we suspect that the diffusion component ($qD_N \nabla n$) is negligible compared to the drift component of the current. Simplifying Eq. (5.4) in accordance with the preceding observations yields

$$J_N = J_{Ny} = q\mu_n n \mathscr{E}_y = -q\mu_n n \frac{dV}{dy} \quad \text{in the conducting channel.} \tag{5.5}$$

μ_n, n, and J_{Ny} in Eq. (5.5) are, of course, all strongly position dependent; J_{Ny} in the MOSFET case is very large for $x = 0^+$ and drops off rapidly as one moves into the semiconductor bulk.

Since there are no current sinks or sources inside the device, the current flowing through any cross-sectional plane within the channel must be equal to I_D. Thus integrating the current density over the cross-sectional area of the conducting channel at an arbitrary point y gives

$$I_D = -\iint J_{Ny}\, dx\, dz = -Z \int_0^{x_c(y)} J_{Ny}\, dx \tag{5.6a}$$

$$= \left(-Z\frac{dV}{dy}\right)\left(-q \int_0^{x_c(y)} \mu_n(x, y) n(x, y)\, dx\right) \tag{5.6b)*}$$

*Generally speaking, V and dV/dy are functions of x. The very narrow extent of the inversion layer dictates, however, that $V \simeq V_{|x=0}$ in the channel region. Thus, in writing down the final form of Eq. (5.6), dV/dy was taken to be constant over the x-width of the channel.

The second bracket on the right-hand side of Eq. (5.6b) is clearly just $\overline{\mu}_n Q_N$ (see Eq. (5.3)) and one can write

$$I_D = -Z\overline{\mu}_n Q_N \frac{dV}{dy} \tag{5.7}$$

Finally, remembering that I_D is independent of y (and $\overline{\mu}_n$ assumed to be independent of y), one can recast Eq. (5.7) into a more useful form by integrating I_D over the length of the channel. Specifically, if V is interpreted to be the channel potential referenced to the channel potential at $y = 0$,

$$\int_0^L I_D\, dy = I_D L = -Z\overline{\mu}_n \int_0^{V_D} Q_N\, dV \tag{5.8}$$

and

$$\boxed{I_D = -\frac{Z\overline{\mu}_n}{L}\int_0^{V_D} Q_N\, dV} \tag{5.9}$$

5.2.3 Square-Law Theory

An analytical expression for the channel charge is required to complete the theoretical analysis. In the simplest of MOSFET theories $Q_N(y)$ is obtained from capacitor-related arguments. Under equilibrium conditions the charge added to the gate of an MOS-C is almost perfectly balanced by increases in the inversion layer charge for V_G's $> V_T$.* Basically, given $V_G > V_T$, charge is simply being added on the two sides of the oxide. In a standard parallel-plate capacitor, however, the charge on the capacitor plates is equal to the capacitance times the voltage drop between the plates. By analogy with the standard capacitor, then, we expect

$$Q_{G|V_G \geq V_T} - Q_{G|V_G = V_T} = -Q_N \tag{5.10a}$$

$$\simeq C_o\left(V_G - \frac{kT}{q}2U_F\right) - C_o\left(V_T - \frac{kT}{q}2U_F\right) \tag{5.10b}$$

or

$$Q_N = -C_o(V_G - V_T) \tag{5.11}$$

Note that the foregoing makes use of the delta-depletion approximation for the voltage on the semiconductor side of the oxide under inversion conditions and is consistent with the fact that $Q_N \simeq 0$ if $V_G = V_T$. Next, the gate region of the transistor is viewed as nothing more than a capacitor with a position dependent voltage of $(kT/q)2U_F + V$ $(0 \leq V \leq V_D)$ on one of the plates. Thus, simply replacing $V_G - (kT/q)2U_F$ in Eq. (5.10b) with

*Assumes an MOS-C with a low density of interfacial traps where changes in Q_{IT} for V_G's $> V_T$ may be neglected.

$V_G - (kT/q)2U_F - V$, one concludes

$$Q_N(y) = -C_o(V_G - V_T - V) \qquad \text{where } V = V(y) \tag{5.12}$$

An explicit I_D–V_D relationship can now be established by substituting the Eq. (5.12) expression for Q_N into Eq. (5.9) and integrating. The result is

$$\boxed{I_D = \frac{Z\bar{\mu}_n C_o}{L}[(V_G - V_T)V_D - V_D^2/2] \qquad \begin{pmatrix} 0 \leq V_D \leq V_{Dsat} \\ V_G \geq V_T \end{pmatrix}} \tag{5.13}$$

The post-pinch-off portion of the characteristics are approximately modeled by setting

$$I_{D|V_D > V_{Dsat}} = I_{D|V_D = V_{Dsat}} \equiv I_{Dsat} \tag{5.14a}$$

or

$$I_{Dsat} = \frac{Z\bar{\mu}_n C_o}{L}[(V_G - V_T)V_{Dsat} - V_{Dsat}^2/2] \tag{5.14b}$$

V_{Dsat} can be eliminated in Eq. (5.14b) by noting $Q_N(L) \to 0$ when $V(L) = V_D \to V_{Dsat}$. Using Eq. (5.12),

$$Q_N(L) = -C_o(V_G - V_T - V_{Dsat}) = 0 \tag{5.15}$$

or

$$\boxed{V_{Dsat} = V_G - V_T} \tag{5.16}$$

and

$$\boxed{I_{Dsat} = \frac{Z\bar{\mu}_n C_o}{2L}(V_G - V_T)^2} \tag{5.17}$$

Neglecting $\bar{\mu}_n$'s dependence on V_G, Eq. (5.17) predicts a saturation drain current which varies as the square of the gate voltage above turn-on, the so-called "square-law" dependence. The reader might recall that a similar square-law relationship, Eq. (1.14), approximately describes the post-pinch-off portion of the J-FET characteristics.

5.2.4 Bulk-Charge Theory

Although appearing very reasonable and sound on first inspection, close scrutiny reveals the square-law theory contains a major flaw. The square-law analysis implicitly assumes that the depletion width for all channel points from the source to the drain remains fixed at W_T even under $V_D \neq 0$ biasing. In reality, as shown in Fig. 5.2(b) to (d), the depletion width widens in progressing from the source to the drain when $V_D \neq 0$. Consequently, the point-to-point variation in the depletion layer or "bulk" charge must be included in any charge balance relationship; that is, changes in the gate charge going down the MOSFET channel are not balanced solely by changes in Q_N, but are in part balanced by changes in the depletion layer charge.

With changes in the depletion width, $W(y)$, taken into account, one more accurately deduces

$$Q_G(y)|_{V_G \geq V_T} - Q_G|_{V_G = V_T} = -Q_N(y) + qN_AW(y) - qN_AW_T \tag{5.18a}$$

$$\simeq C_o\left(V_G - V - \frac{kT}{q}2U_F\right) - C_o\left(V_T - \frac{kT}{q}2U_F\right) \tag{5.18b}$$

or

$$Q_N(y) = -C_o(V_G - V_T - V) + qN_AW_T\left[\frac{W(y)}{W_T} - 1\right] \tag{5.19}$$

where, making use of the delta-depletion results in Chapter 2,

$$W(y) = \left[\frac{2K_S\varepsilon_0}{qN_A}\left(\frac{kT}{q}2U_F + V\right)\right]^{1/2} \tag{5.20}$$

$$W_T = \left[\frac{2K_S\varepsilon_0}{qN_A}\frac{kT}{q}2U_F\right]^{1/2} \tag{5.21}$$

Thus, combining Eqs. (5.19) to (5.21) and introducing

$$V_W \equiv \frac{qN_AW_T}{C_o} \tag{5.22}$$

$$\phi_F \equiv \frac{kT}{q}U_F \tag{5.23}$$

one obtains the bulk-charge theory analogue of Eq. (5.12), namely,

$$Q_N(y) = -C_o\left[V_G - V_T - V - V_W\left(\sqrt{1 + V/2\phi_F} - 1\right)\right] \tag{5.24}$$

The predicted I_D–V_D relationship based on the bulk-charge formulation is next readily obtained by substituting Eq. (5.24) into Eq. (5.9) and integrating. The end result is

$$\boxed{I_D = \frac{Z\bar{\mu}_nC_o}{L}\left\{(V_G - V_T)V_D - \frac{V_D^2}{2} - \frac{4}{3}V_W\phi_F\left[\left(1 + \frac{V_D}{2\phi_F}\right)^{3/2} - \left(1 + \frac{3V_D}{4\phi_F}\right)\right]\right\}}$$

$$\text{for } 0 \leq V_D \leq V_{Dsat} \text{ and } V_G \geq V_T \tag{5.25}$$

As in previous analyses the post-pinch-off portion of the characteristics are approximately modeled by setting I_D evaluated at $V_D > V_{Dsat}$ equal to I_D at $V_D = V_{Dsat}$. Likewise, an expression for V_{Dsat} can be obtained by noting $Q_N(y)|_{y=L} \to 0$ in Eq. (5.24) when

$V(L) = V_D \rightarrow V_{Dsat}$. One finds (after a nontrivial amount of algebraic manipulation)

$$V_{Dsat} = V_G - V_T - V_W\left\{\left[\frac{V_G - V_T}{2\phi_F} + \left(1 + \frac{V_W}{4\phi_F}\right)^2\right]^{1/2} - \left(1 + \frac{V_W}{4\phi_F}\right)\right\} \quad (5.26)$$

Having concluded the mathematical development, it is time to step back, examine the results, and make appropriate comments. First of all, it should be recognized that the primary asset of the square-law theory is its simplicity. General trends, basic interrelationships, etc., can be established using the square-law formulation without an excessive amount of mathematical entanglement. The bulk-charge theory, on the other hand, is in excellent agreement with the experimental characteristics derived from long-channel MOSFET's. Secondly, it should be noted that, although Eqs. (5.25) and (5.26) are decidedly more complex than their square-law analogues, the added terms, the terms not appearing in Eqs. (5.13) and (5.16), respectively, are always negative and act primarily to reduce I_D and V_{Dsat} for a given set of operational conditions. Figure 5.8, which compares the two theories, confirms the foregoing observation and also illustrates another well-known property—the accuracy of the square-law theory improves as the substrate doping is decreased. In fact, the bulk-charge theory mathematically reduces to the square-law theory as N_A (or N_D) $\rightarrow 0$ and $x_o \rightarrow 0$. Finally, the reader should be made aware of the implicit use of the gradual channel approximation in the MOSFET analysis. The assumed balancing of the gate and semiconductor charges at all points along the MOSFET channel when $V_D \neq 0$ is equivalent to invoking the gradual channel approximation, a common

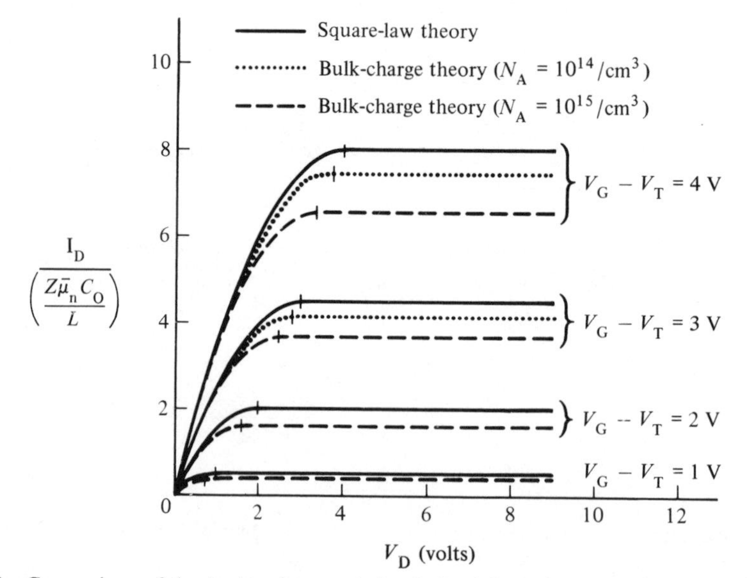

Fig. 5.8 Comparison of the I_D–V_D characteristics derived from the square-law and bulk-charge theories. The bulk-charge curves were computed assuming $x_o = 0.1\mu$ and $T = 23°C$.

approximation in FET work which was initially introduced and described in the J-FET analysis of Section 1.3.

5.3 THRESHOLD CONSIDERATIONS

Perhaps because of the method of presentation, fledgling device analysts are sometimes under the impression that the nonidealities discussed in relationship to the MOS-C do not affect the MOSFET. This impression is, of course, totally false. Sodium ions in the oxide, for example, can cause the threshold or turn-on voltage for both n- and p-channel devices to occur at large negative gate biases. Moreover, movement of the ions within the oxide can cause the drain current observed at a given bias point to drift as a function of time. (The threshold voltage changes with time corresponding to the voltage translation of the C–V characteristics in the MOS-C analysis.) If present in large densities the interfacial traps which "spread out" the MOS-C C–V characteristics can likewise increase the change in gate voltage required to achieve a desired ΔI_D at a given drain voltage. In other words, interfacial traps can reduce the "gain" ($\Delta I_D/\Delta V_{G|V_D}$) of a transistor. While it is true that both the mobile ion and interfacial trap problems were minimized early in MOSFET development, the remaining nonidealities, primarily through their effect on V_T, have had a large (incredibly large) impact on fabrication technology, device design, and modes of operation. In this section we will examine a number of related items which generally fall under the heading of threshold considerations.

5.3.1 Threshold Voltage Relationships

An expression for the threshold voltage V_T' exhibited by an ideal device is readily established using the delta-depletion formulation and relevant relationships previously presented in Section 3.2. Specifically, combining Eqs. (3.7) to (3.9) and noting $V_G' = V_T'$ when $U_S = 2U_F$, one obtains

$$V_T' = \frac{kT}{q}2U_F + \frac{q(N_A - N_D)}{K_S\varepsilon_0}x_o'\left[\frac{2K_S\varepsilon_0}{q(N_A - N_D)}\frac{kT}{q}2U_F\right]^{1/2} \tag{5.27}$$

or

$$V_T' = \begin{cases} 2\phi_F + \dfrac{1}{C_o}\sqrt{4qN_AK_S\varepsilon_0\phi_F} & \text{for } n\text{-channel } (p\text{-bulk}) \text{ devices} \quad (5.28a) \\[2ex] 2\phi_F - \dfrac{1}{C_o}\sqrt{4qN_DK_S\varepsilon_0(-\phi_F)} & \text{for } p\text{-channel } (n\text{-bulk}) \text{ devices} \quad (5.28b) \end{cases}$$

where, as previously defined, $\phi_F = (kT/q)U_F$.

An expression for the threshold voltage V_T exhibited by a real device is next established by simply evaluating Eq. (4.16) at the $U_S = 2U_F$ point; i.e.,

$$V_T = V_T' + \phi_{MS} - \frac{Q_F}{C_o} - \frac{Q_M\gamma_M}{C_o} - \frac{Q_{IT}(2U_F)}{C_o} \tag{5.29}$$

Moreover, note that evaluating Eq. (4.16) at the Flat Band ($U_S = 0$) point yields

$$V_{FB} \equiv V_{G|U_S=0} = \phi_{MS} - \frac{Q_F}{C_o} - \frac{Q_M \gamma_M}{C_o} - \frac{Q_{IT}(0)}{C_o} \tag{5.30}$$

Thus, if $Q_{IT}(2U_F)/C_o \simeq Q_{IT}(0)/C_o$, a reasonably good approximation in well-made devices, one can also write

$$\boxed{V_T = V_{FB} + V_T'} \tag{5.31}$$

5.3.2 Threshold, Terminology, and Technology

As a lead into the discussion let us perform a simple threshold voltage computation employing relationships developed in the preceding subsection. Suppose the gate material is Al, the Si surface orientation is (111), $T = 23°C$, $x_o = 0.1\ \mu$, $N_A = 10^{15}/cm^3$, $Q_F/q = 2 \times 10^{11}/cm^2$, $Q_M = 0$, and $Q_{IT} = 0$. For the given n-channel device one computes $\phi_{MS} = -0.98$ V, $-Q_F/C_o = -0.93$ V, $V_{FB} = -1.91$ V, $V_T' = 1.00$ V and $V_T = -0.91$ V. Observe: whereas V_T' is positive, as expected, nonidealities of a very realistic magnitude cause V_T to be negative. Since an n-channel device turns-on for gate voltages $V_G > V_T$, the device in question is already "on" at a gate bias of zero volts. Actually, negative biases must be applied to turn the device off! For a p-channel device with identical parameters (except, of course, for an N_D doped substrate) one obtains a $V_T' = -1.00$ V, $V_{FB} = -1.91$ V, and $V_T = -2.91$ V. In the p-channel case the considered nonidealities merely increase the negative voltage required to achieve turn-on.

When a MOSFET is "on" at $V_G = 0$ V, the transistor is referred to as a *depletion mode* device; when a MOSFET is "off" at $V_G = 0$ V, it is called an *enhancement mode* device. Routinely fabricated p-channel MOSFET's constructed in the standard configuration are ideally and practically enhancement mode devices. n-channel MOSFET's are also ideally enhancement mode devices. However, because nonidealities tend to shift the threshold voltage toward negative biases in the manner indicated in our sample calculation, early n-channel MOS transistors were typically of the depletion mode type. Up until ~1977 this difference in behavior led to the total dominance of *PMOS* technology over *NMOS* technology; that is, IC's incorporating p-channel MOSFET's dominated the commercial marketplace. Subsequently, as explained under the heading of threshold adjustment, NMOS, which is to be preferred because of the greater mobility of electrons compared to holes, benefited from technological innovations widely implemented in the late 1970s and is now incorporated in the majority of newly designed IC's.

While on the topic of the threshold voltage in practical devices, it is relevant to note that the inversion threshold of regions adjacent to the device proper is also of concern. Consider, for example, the unmetallized region between the two n-channel MOS transistors pictured in Fig. 5.9(a). If the potential at the unmetallized outer oxide surface is assumed to be zero (normally a fairly reasonable assumption) and if the threshold voltage for the n-channel transistors is negative, then the intermediate region between the two transistors will be inverted. In other words, a conducting path, a pseudo-channel, will exist between the transistors. This undesirable condition was another nuisance in early

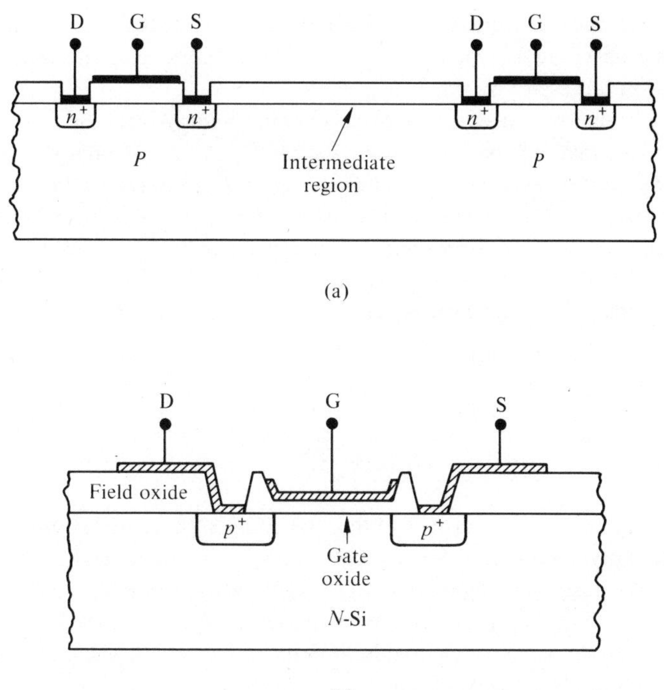

Fig. 5.9 (a) Visualization of the intermediate region between two MOSFET's. (b) Identification of the gate-oxide and field-oxide regions in practical MOSFET structures.

NMOS technology, where, as already noted, nonidealities tended to invert the surface of the semiconductor in the absence of an applied gate bias. Unless special precautions are taken, unwanted pseudo-channels between devices can also arise in both n- and p-channel IC's from the potential applied to the metal overlays supplying the gate and drain biases. To avoid this problem the oxide in the nongated portions of the IC, referred to as the *field-oxide*, is typically much thicker than the *gate-oxide* in the active regions of the structure (see Fig. 5.9(b)). The idea behind the use of the thicker oxide can be understood by referring to Eqs. (5.28) and (5.30). Both V_T' and V_{FB} contain terms which are proportional to $1/C_o = x_o/K_O\varepsilon_0$. Thus employing an x_o (field-oxide) $\gg x_o$ (gate-oxide) increases $|V_T|$ in the field-oxide areas relative to the gated areas in PMOS (and modern NMOS) structures. Inversion of the field-oxide regions is thereby avoided at potentials normally required for IC operation.

5.3.3 Threshold Adjustment

Several physical factors affect the threshold voltage and can therefore be used to vary the V_T actually exhibited by a given MOSFET. We have, in fact, already cited the adjustment of V_T through a variation of the oxide thickness. Obviously, the substrate doping can also be varied to increase or decrease the threshold voltage. However, although strongly influencing the observed V_T value, the gate-oxide thickness and substrate doping are predetermined in large part by other design restraints.

Other factors that play a significant role in determining V_T are the substrate surface orientation and the material used in forming the MOS gate. As first noted in Section 4.3, the Q_F in MOS devices constructed on (100) surfaces is ~3 times smaller than the Q_F in devices constructed on (111) surfaces. The use of (100) substrates therefore reduces the ΔV_G associated with the fixed oxide charge. The use of a polysilicon instead of an Al gate, on the other hand, makes ϕ_{MS} less negative or even positive. Given a polysilicon gate the effective "metal" workfunction becomes

$$\text{“}\Phi_M\text{”} = \chi_{Si} + (E_c - E_F)_{\text{poly-Si}} \tag{5.32}$$

and

$$\phi_{MS} = \frac{1}{q}[(E_c - E_F)_{\text{poly-Si}} - (E_c - E_F)_{\infty,\ \text{crystalline-Si}}] \tag{5.33}$$

If we redo the calculation performed in the preceding subsection assuming a typically doped p-type polysilicon gate where $E_F \simeq E_v$, we obtain a $\phi_{MS} = +0.26$ V and $V_{FB} \rightarrow -0.67$ V. If the substrate orientation is also changed from (111) to (100) causing a threefold reduction in Q_F, V_{FB} is further increased to $V_{FB} = -0.05$ V. *Note*: V_T now becomes $V_T = +0.95$ V. Thus positive NMOS thresholds are possible in (100)-oriented structures incorporating polysilicon gates.

Although the foregoing calculation shows positive threshold voltages are possible, actual structures may be only nominally positive. For various reasons a larger threshold voltage may be desired, or one may desire to modify the threshold attainable in a PMOS structure, or tailoring of the threshold for both n- and p-channel devices on the same IC chip may be required. For a number of reasons, it is very desirable to have a flexible threshold adjustment process where V_T can be controlled essentially at will. In modern device processing this is accomplished through the use of *ion implantation*.

The general ion implantation process was described in Section 1.1 of Volume II. To adjust the threshold voltage, a relatively small, precisely controlled number of either boron or phosphorus ions is implanted into the near-surface region of the semiconductor. When the MOS structure is depletion or inversion biased, the implanted dopant adds to the exposed dopant-ion charge near the oxide–semiconductor interface and thereby translates the V_T exhibited by the structure. The implantation of boron causes a positive shift in the threshold voltage; phosphorus implantation causes a negative voltage shift. For shallow implants the procedure may be viewed to first-order as placing an additional "fixed" charge at the oxide–semiconductor interface. If N_I is the number of implanted ions/cm^2 and $Q_I = \pm qN_I$ is the implant-related donor (+) or acceptor (−) charge/cm^2 at the oxide–semiconductor interface, then, by analogy with the fixed charge analysis of Section 4.3 (Volume IV),

$$\Delta V_{G|\text{ions}} = -\frac{Q_I}{C_o} \tag{5.34}$$

Assuming, for example, an $N_I = 5 \times 10^{11}$ boron ions/cm^2 and an $x_o = 0.1\ \mu$, one computes a threshold adjustment of $+2.32$ V.

5.3.4 Back Biasing

Reverse biasing the back contact or bulk of a MOS transistor relative to the source is another method which has been employed to adjust the threshold potential. This electrical method of adjustment, which predates ion implantation, makes use of the so-called *body effect* or substrate-bias effect.

To explain the effect let us consider the n-channel MOSFET shown in Fig. 5.10(a). If the back-to-source potential difference (V_{BS}) is zero, inversion occurs of course when the voltage drop across the semiconductor $[(kT/q)U_S]$ equals $2\phi_F$ as pictured in Fig. 5.10(b). If $V_{BS} < 0$, the semiconductor still attempts to invert when $(kT/q)U_S$ reaches $2\phi_F$. However, with $V_{BS} < 0$ any inversion layer carriers which do appear at the semiconductor surface migrate laterally into the source and drain because these regions are at a lower potential. Not under $(kT/q)U_S = 2\phi_F - V_{BS}$, as pictured in Fig. 5.10(c), will the surface invert and normal transistor action begin. In essence, back

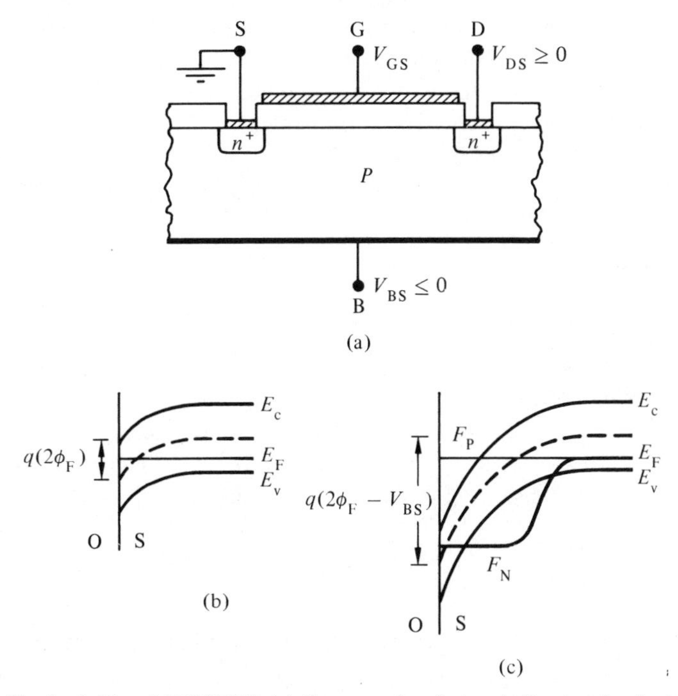

Fig. 5.10 The back-biased MOSFET. (a) Cross-sectional view indicating the double subscripted voltage variables used in the analysis. Also shown are the semiconductor energy band diagrams corresponding to the onset of inversion when (b) $V_{BS} = 0$ and (c) $V_{BS} < 0$.

biasing changes the inversion point in the semiconductor from $2\phi_F$ to $2\phi_F - V_{BS}$. The ideal device threshold potential given by Eq. (5.28a) is in turn modified to

$$V'_{GB|\text{at threshold}} = 2\phi_F - V_{BS} + \frac{1}{C_o}\sqrt{2qN_AK_S\varepsilon_0(2\phi_F - V_{BS})} \quad \text{for } n\text{-channel devices } (V_{BS} < 0) \tag{5.35}$$

Since $V'_{GB|\text{at threshold}} = V'_{GS|\text{at threshold}} - V_{BS}$, we can alternatively write

$$V'_{GS|\text{at threshold}} = \begin{cases} 2\phi_F + \dfrac{1}{C_o}\sqrt{2qN_AK_S\varepsilon_0(2\phi_F - V_{BS})} & \text{for } n\text{-channel devices } (V_{BS} < 0) \quad (5.36a) \\ 2\phi_F - \dfrac{1}{C_o}\sqrt{2qN_DK_S\varepsilon_0(V_{BS} - 2\phi_F)} & \text{for } p\text{-channel devices } (V_{BS} > 0) \quad (5.36b) \end{cases}$$

Having established Eq. (5.36), we make the following observations concerning back biasing or the body effect: (1) Back biasing always increases the magnitude of the ideal device threshold voltage. It therefore makes the p-channel threshold of actual devices more negative and the n-channel threshold more positive—it cannot be used to reduce the negative threshold of a p-channel MOSFET. (2) The current–voltage relationships developed in Section 5.2 are still valid when $V_{BS} \neq 0$ provided $2\phi_F \to 2\phi_F - V_{BS}$, $V_G \to V_{GS}$, $V_D \to V_{DS}$, and V_T is interpreted as $V_{GS|\text{at threshold}}$. (3) Care must be exercised in describing back-biased structures to properly identify voltage differences through the use of double-subscripted voltage variables. The use of double subscripts is, in fact, standard practice in circuit oriented MOS work. However, for convenience and simplicity of notation, the single subscripts (with the back contact assumed to be at ground potential) are routinely employed in works primarily concerned with MOS device physics.

5.4 ac RESPONSE

5.4.1 Small Signal Equivalent Circuits

If one examines the development leading to the low-frequency circuit for the J-FET it becomes immediately obvious that, with very little modification, the Section 1.4 arguments are equally valid for the MOSFET. Without further explanation one can therefore assert that the low-frequency ac response of the MOSFET is characterized by the small signal equivalent circuit displayed in Fig. 5.11(a), where

$$g_d \equiv \left.\frac{\partial I_D}{\partial V_D}\right|_{V_G=\text{constant}} \quad \text{the drain or channel conductance} \tag{5.37a}$$

$$g_m \equiv \left.\frac{\partial I_D}{\partial V_G}\right|_{V_D=\text{constant}} \quad \text{transconductance or mutual conductance} \tag{5.37b}$$

Explicit g_d and g_m relationships obtained by direct differentiation of Eqs. (5.13), (5.17), and (5.25) using the Eq. (5.37) definitions are catalogued in Table 5.1.

Table 5.1 MOSFET Small Signal Parameters.*

	Below Pinch-Off ($V_D \le V_{Dsat}$)	Post-Pinch-Off ($V_D > V_{Dsat}$)
Square law	$g_d = \frac{Z\bar{\mu}_n C_o}{L}(V_G - V_T - V_D)$	$g_d = 0$
Bulk charge	$g_d = \frac{Z\bar{\mu}_n C_o}{L}[V_G - V_T - V_D - V_W(\sqrt{1 + V_D/2\phi_F} - 1)]$	$g_d = 0$
Square law	$g_m = \frac{Z\bar{\mu}_n C_o}{L} V_D$	$g_m = \frac{Z\bar{\mu}_n C_o}{L}(V_G - V_T)$
Bulk charge	$g_m = \frac{Z\bar{\mu}_n C_o}{L} V_D$	$g_m = \frac{Z\bar{\mu}_n C_o}{L} V_{Dsat}$ with V_{Dsat} per Eq. (5.26)

*Entries in the table were obtained by direct differentiation of Eqs. (5.13), (5.17), and (5.25). The variation of $\bar{\mu}_n$ with V_G was neglected in establishing the g_m expressions.

At the higher operational frequencies often encountered in practical applications, the Fig. 5.11(a) circuit must be modified to take into account capacitive coupling between the device terminals. The required modification is shown in Fig. 5.11(b). A capacitor between the drain and source terminals at the output has been omitted in Fig. 5.11(b) because the drain-to-source capacitance is typically negligible. C_{gd}, which provides undesirable feedback between the input and output, is associated in large part with the so-called overlap capacitance—the capacitance resulting from the portion of the gate that

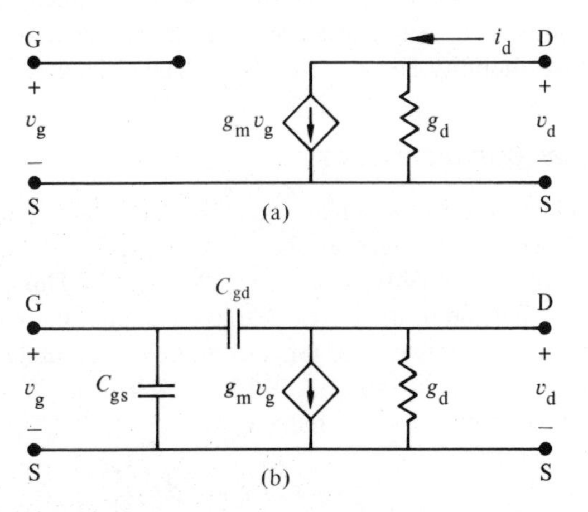

Fig. 5.11 Small signal equivalent circuits characterizing the (a) low-frequency and (b) high-frequency ac response of the MOSFET.

overlaps the drain island. The overlap capacitance is minimized by forming a thicker oxide in the overlap region or through the use of self-aligned gate procedures. In the self-aligned gate fabrication process the MOSFET gate, necessarily poly-Si or a refractory metal which can withstand high temperature processing, is deposited first, and the source and drain islands are subsequently formed abutting the gate by diffusion or ion implantation. The remaining capacitor shown in Fig. 5.11(b), C_{gs}, is of course associated primarily with the capacitance of the MOS gate.

5.4.2 Cutoff Frequency

Given the small signal equivalent circuit of Fig. 5.11(b) it is possible to estimate the maximum operating frequency or cutoff frequency of an MOS transistor. Let f_{max} be defined as the frequency where the MOSFET is no longer amplifying the input signal under optimum conditions; i.e., the frequency where the absolute value of the output current to input current ratio is unity when the output of the transistor is short-circuited. By inspection, the input current with the output short-circuited is

$$i_{in} = j\omega(C_{gs} + C_{gd})v_g \simeq j(2\pi f)C_O v_g \qquad (j = \sqrt{-1}) \tag{5.38}$$

The output current is

$$i_{out} = g_m v_g \tag{5.39}$$

Thus, setting $|i_{out}/i_{in}| = 1$ and solving for $f = f_{max}$, one obtains

$$f_{max} = \frac{g_m}{2\pi C_O} = \frac{\bar{\mu}_n V_D}{2\pi L^2} \qquad \text{if } V_D \le V_{Dsat} \tag{5.40}$$

The latter form of Eq. (5.40) was established using the below pinch-off g_m entry in Table 5.1. The important point to note is that for high frequency operation it is desirable to maximize the effective mobility and minimize the channel length.

5.4.3 Small Signal Characteristics

Representative sketches of selected small signal characteristics which have received special attention in the device literature are shown in Fig. 5.12. g_d versus V_G with $V_D = 0$ has been used to obtain a reasonably accurate estimate of V_T. This is accomplished by extrapolating the linear portion of the g_d–V_G characteristic into the V_G axis and equating the voltage intercept to V_T. The basis for this procedure can be understood by referring to the below pinch-off g_d entries in Table 5.1. With $V_D = 0$ the drain conductance in both the square-law and bulk-charge theories reduces to

$$g_d = \frac{Z\bar{\mu}_n C_o}{L}(V_G - V_T) \qquad (V_D = 0) \tag{5.41}$$

To first order, then, g_d is predicted to be a linear function of V_G, going to zero when $V_G = V_T$. The experimental characteristic does not completely vanish at $V_G = V_T$ because

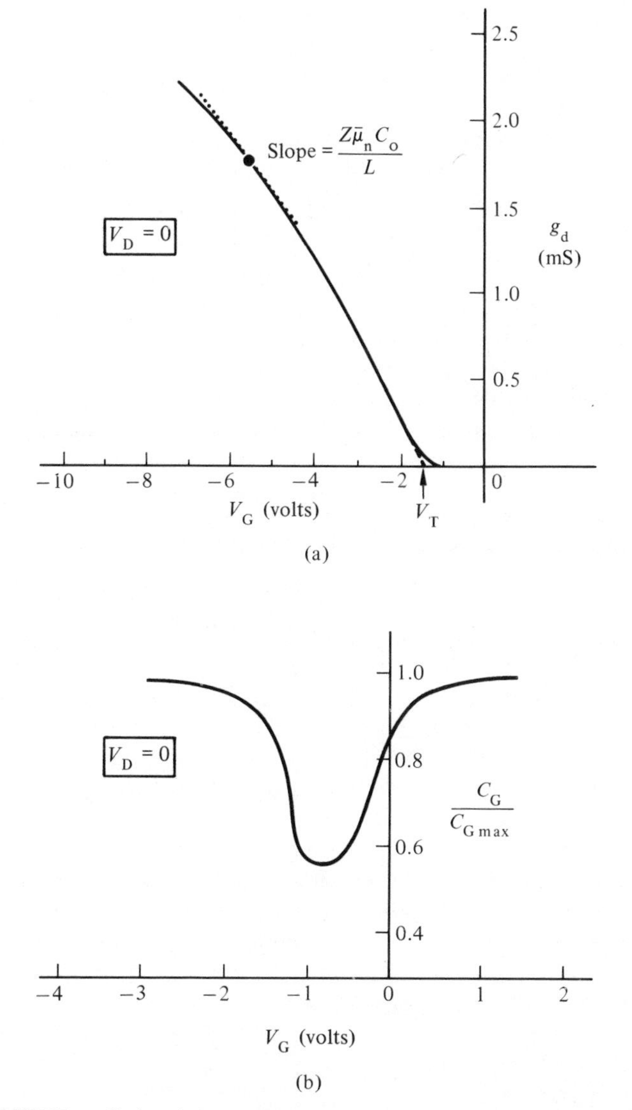

Fig. 5.12 MOSFET small signal characteristics. (a) g_d versus V_G with $V_D = 0$; (b) C_G versus V_G with $V_D = 0$.

of a small, nonzero, minority carrier concentration which actually exists at the depletion–inversion transition point and which is neglected in both the square-law and bulk-charge theories. The g_d versus V_G characteristic with $V_D = 0$ also underlies the most widely employed method for measuring the effective mobility. Since g_d is directly proportional to $\overline{\mu}_n$ according to Eq. (5.41), $\overline{\mu}_n$ versus V_G can be deduced readily from the slope of the g_d–V_G characteristic. This mobility measurement method is accurate provided the device contains a low density of interfacial traps. A moderate or large density of interface states would "spread out" the g_d–V_G characteristic and yield a fallaciously low value for $\overline{\mu}_n$.

The second characteristic in Fig. 5.12 typifies the gate capacitance versus V_G dependence derived from the MOSFET when the drain is grounded. The MOSFET C_G–V_G ($V_D = 0$) characteristic can be used as a diagnostic tool in much the same manner as the MOS-C C–V_G characteristic and is in fact modeled to first-order by the low-frequency MOS-C C–V_G theory. Note, however, that a low-frequency type characteristic is observed even though the MOSFET measurement frequency may be as high as 1 MHz. A low frequency characteristic is obtained because the source and drain islands supply the minority carriers required for the structure to quasi-statically follow the ac fluctuations in the gate potential under dc inversion biasing. Minority carriers merely use the surface channel to flow laterally into and out of the MOS gate area in response to the applied ac signal.

5.5 SUMMARY AND CONCLUDING COMMENTS

In this chapter we presented relevant information about the MOSFET and examined the basic principles of MOSFET operation. Although the reader may have grown tired of the repeated allusions to the J-FET, one last overall comparison of the two devices is most helpful in reviewing the qualitative theory of operation for the MOSFET. Externally the J-FET and MOSFET yield similar electrical characteristics and even appear similar physically, with the terminal leads being designated the source, drain, and gate. The gate voltage in both FET's determines the maximum conductance of the internal channel between the source and drain. The drain voltage in both devices initiates current flow between the source and drain terminals. The current flow is proportional to V_D at low drain voltages, slopes-over due to channel narrowing as V_D is increased, and eventually saturates once the internal channel vanishes or pinches off near the drain. Major differences between the two devices lie in the nature of the conducting channel and the substructure used to modulate the channel conductance. The J-FET channel is a narrow piece of bulk material; the usual MOSFET channel is a surface inversion layer which (in ideal devices) is created by the applied gate voltage. Manipulation of the channel conductance is accomplished by reverse biasing a *p*-*n* junction in the J-FET; the MOSFET channel conductance is manipulated by the bias applied to an MOS structure.

The quantitative analysis of the MOSFET's dc characteristics also closely parallels the J-FET analysis but is subject to two complicating factors not encountered in the earlier analysis. First of all, carriers in a surface channel experience motion-impeding collisions with the Si surface which lower the mobility of the carriers and necessitate the introduction of an effective carrier mobility. Secondly, the carrier concentration in the surface channel is a strong function of position, dropping off rapidly as one proceeds into the semiconductor bulk. The first-order results for the MOSFET I_D–V_D relationship, however, are surprisingly simple. The results of the first-order theory, referred to herein as the square-law theory, are contained in Eqs. (5.13), (5.16), and (5.17). A second formulation, the bulk-charge theory, was also considered and led to the results contained in Eqs. (5.25) and (5.26). The latter theory provides a more accurate representation of reality at the expense of increased complexity.

The final sections of the chapter were addressed to threshold considerations and the ac response of the MOSFET. The threshold voltage is an important parameter in practical applications and is carefully adjusted in modern devices through the use of ion implantation. Back biasing, making use of the body effect, has also been used to achieve a degree of threshold control. The inversion threshold in regions outside of the gated areas in IC's is likewise of concern and explains certain design features such as the thicker field-oxide in the nongated regions. The ac response of the MOSFET is best summarized by simply referring to Figs. 5.11 and 5.12. Based on the cutoff frequency calculation one might lastly add that a short channel length is a necessary prerequisite for high-frequency, high-speed operation.

The reader should be cautioned that the MOSFET considerations herein are only an introduction, a first go-around, so to speak. Nothing has been said about more exacting theories, post-pinch-off models, and computer generated results. Also, the widespread utilization of the MOSFET has led to many structural variations, each of which requires special considerations and a modified analysis. The unrelenting push for greater and greater miniaturization to achieve higher operating speeds and increased packing densities has, moreover, led to major deviations in operation from that described herein. In 1965 a "short" channel transistor had an $L \sim 1$ mil $\simeq 25\ \mu$. 1980 versions of the MOSFET boast design features with 3 μ to 5 μ line widths and effective channel lengths down to $\sim 1\ \mu$. Even submicron channel length devices have been fabricated in research laboratories and will reach production-line status in the near future. In such devices one must worry about punch-through between the source and drain, drain voltages affecting the turn-on voltage, and other previously unimagined effects. We make these comments not to scare the reader but merely to point out the existence of a vast pool of additional information about the MOSFET.

PROBLEMS

5.1 Answer the following questions as concisely as possible.

(a) Precisely what is the "channel" in MOSFET terminology?

(b) Define "threshold voltage."

(c) Sketch an outline of the inversion layer and depletion region inside a MOSFET biased at the pinch-off point.

(d) Why does the V_G = constant drain current in short channel devices increase somewhat with increasing $V_D > V_{Dsat}$?

(e) Why is the mobility in the surface channel of a MOSFET different from the carrier mobility in the semiconductor bulk?

(f) Explain what is meant by the term "depletion mode" transistor.

(g) What is the difference between the "field-oxide" and the "gate-oxide"?

(h) Precisely what is the "body effect"?

(i) What is the mathematical definition of the drain conductance? the transconductance?

(j) Why is the observed MOSFET C_G–V_G ($V_D = 0$) curve typically a low-frequency characteristic even at a measurement frequency of 1 MHz?

5.2 An n-channel MOSFET maintained at room temperature is characterized by the following parameters: $Z = 50\ \mu$, $L = 5\ \mu$, $x_o = 0.1\ \mu$, $N_A = 2 \times 10^{15}/\text{cm}^3$, $\bar{\mu}_n = 650\ \text{cm}^2/\text{V-sec}$ (assumed independent of V_G) and $V_T = 1$ V (adjusted by ion implantation). Also $K_O = 3.9$, $K_S = 11.8$, $kT/q = 0.0255$ V, and $n_i = 8.6 \times 10^9/\text{cm}^3$. Determine:

(a) I_{Dsat} (square-law theory) if $V_G = 3$ V;

(b) V_{Dsat} (bulk-charge theory) if $V_G = 3$ V;

(c) I_{Dsat} (bulk-charge theory) if $V_G = 3$ V;

(d) g_d if $V_G = 3$ V and $V_D = 0$;

(e) g_m (square-law theory) if $V_G = 3$ V and $V_D = 3$ V;

(f) f_{max} if $V_G = 3$ V and $V_D = 1$ V.

5.3 Suppose the gate of an n-channel MOSFET is connected to the drain making $V_G = V_D$ while the source and back are grounded. Utilizing the square-law results

(a) Sketch I_D vs. V_D ($V_D \geq 0$) if $V_T > 0$.

(b) Sketch I_D vs. V_D ($V_D \geq 0$) if $V_T < 0$.

Place the part (a) and (b) answers on the same set of coordinates and record equations, reasoning, etc. that led to your answers.

5.4 Before bias–temperature stressing a certain MOSFET the I_D–V_D characteristics are as sketched in Fig. P5.4a. Before and *after* bias–temperature stressing using a positive bias the g_d–V_G ($V_D = 0$) characteristics of the same device are as shown in Fig. P5.4b.

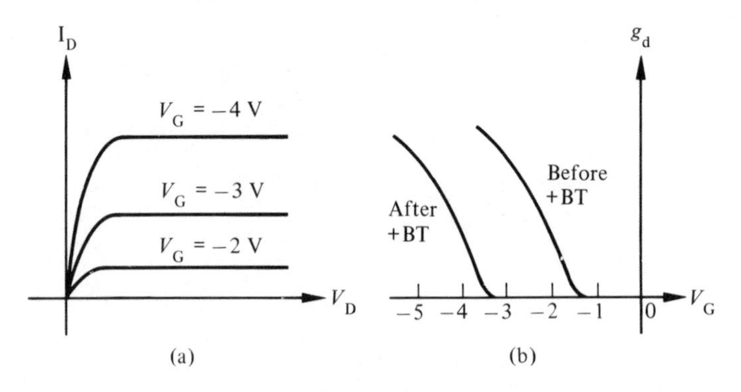

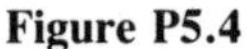
Figure P5.4

(a) What was the probable cause of the shift in the g_d–V_G characteristic after positive bias–temperature stressing?

(b) Sketch the I_D–V_D characteristics *after* positive bias–temperature stressing for gate voltages of $V_G = -2$, -3, and -4 V. Also record your reasoning.

5.5 The measured C_G–V_G ($V_D = 0$) characteristic of a particular n-channel MOSFET is shown in Fig. P5.5

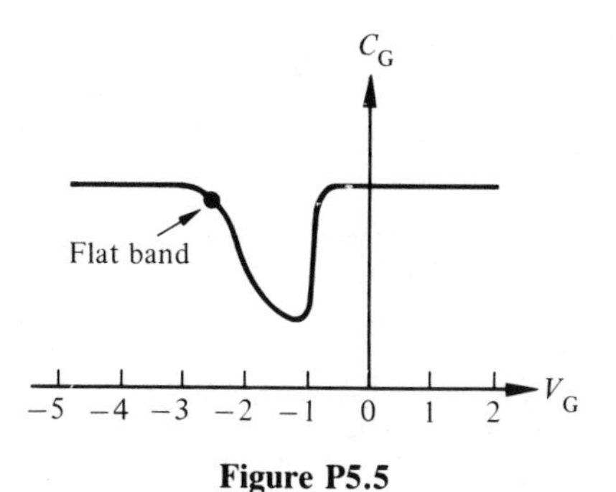

Figure P5.5

(a) What is the threshold voltage, V_T, for the transistor? Explain how you deduced your numerical answer.

(b) Is the MOSFET a depletion mode or enhancement mode device? Explain.

(c) Sketch the I_D–V_D characteristics for the device specifically labeling those characteristics corresponding to $V_G = -2$, -1, 0, 1, and 2 volts.

(d) Given $\phi_{MS} = -1$ V, $Q_{IT} = 0$, and the fact that the device is stable under bias–temperature stressing, how do you explain the observed flat band voltage of -2.5 V?

5.6 *Given* an Al–SiO_2–Si MOSFET, $T = 23°C$ ($kT/q = 0.0255$ V, $E_G = 1.12$ eV, $n_i = 8.6 \times 10^9/cm^3$), a Si substrate doping of $N_D = 5 \times 10^{15}/cm^3$, $x_o = 0.1\ \mu$, $K_O = 3.9$, $K_S = 11.8$, $Q_F/q = 2 \times 10^{11}/cm^2$, no interfacial traps and no mobile ions in the oxide. *Determine* the boron ions/cm² (N_I) which must be implanted into the structure to achieve a $V_T = -1$ V threshold voltage. You may assume the implanted ions create an added negative charge at the Si–SiO_2 interface.

5.7 In the text derivation of the I_D–V_D relationship it was implicitly assumed the MOSFET possessed the usual linear geometry where the gated area is a rectangle of length L and width Z. It is also possible to build MOSFET's with circular geometry as pictured (top view) in Fig. P5.7.

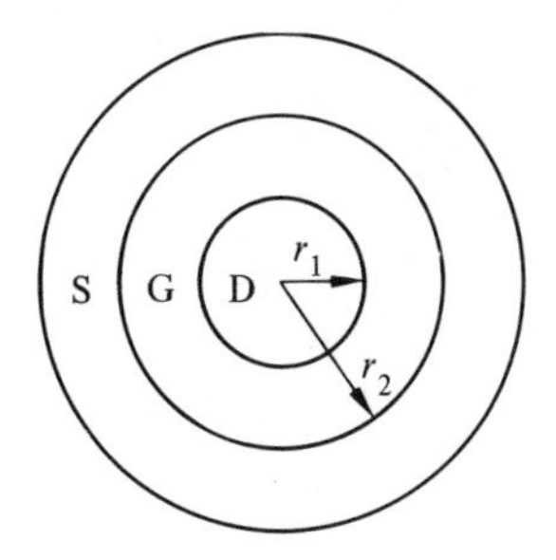

Figure P5.7

(a) If r_1 and r_2 are the inside and outside diameters of the gated area, show that in the square-law formulation one obtains

$$I_D = \frac{2\pi}{\ln(r_2/r_1)}\bar{\mu}_n C_o[(V_G - V_T)V_D - V_D^2/2]$$

for below pinch-off operation of a MOSFET with circular geometry. To derive the above result use cylindrical coordinates (r, θ, z) and appropriately modify Eqs. (5.5) through (5.13).

(b) Setting $r_2 = r_1 + L$ and $Z = 2\pi r_1$ show that the part (a) result reduces to the linear geometry result, Eq. (5.13), in the limit where $L/r_1 \ll 1$.

5.8 Making free use of the square-law entries in Table 5.1, ignoring the variation of $\bar{\mu}_n$ with V_G, and *employing only one set of coordinates per each part of the problem*, draw

(a) $g_m/(Z\bar{\mu}_n C_o/L)$ vs. V_D $(0 \leq V_D \leq 5$ V) when $V_G - V_T = 1, 2,$ and 3 V.

(b) $g_m/(Z\bar{\mu}_n C_o/L)$ vs. V_G $(0 \leq V_G \leq 5$ V) if $V_T = 1$ V and $V_D = 1, 2,$ and 3 V.

(c) $g_d/(Z\bar{\mu}_n C_o/L)$ vs. V_G $(0 \leq V_G \leq 5$ V) if $V_T = 1$ V and $V_D = 0, 1,$ and 2 V.

(d) $g_d/(Z\bar{\mu}_n C_o/L)$ vs. V_D $(0 \leq V_D \leq 5$ V) when $V_G - V_T = 1, 2,$ and 3 V.

Suggested Readings

General Comments: Listed below are four texts which, in the author's opinion, best supplement or enhance the material presented in this volume. The cited texts, it should be noted, contain additional references that might be consulted by the reader.

1. R. A. Colclaser, *Microelectronics: Processing and Device Design*. New York: Wiley, 1980. This up-to-date work may be helpful in answering questions concerning fabricational details and device design. Chapter 8 is specifically devoted to devices for MOS integrated circuits.

2. A. S. Grove, *Physics and Technology of Semiconductor Devices*. New York: Wiley, 1967. See Chapters 9, 11, and 12. Grove gives an authoritative coverage of field effect device fundamentals. This work is perhaps a bit dated in spots, but is still very useful as an aid in understanding the subject matter.

3. B. G. Streetman, *Solid State Electronic Devices*. 2nd edition. Englewood Cliffs, N. J. Prentice-Hall, 1980. See Chapter 8. The book by Streetman is a good general-purpose text, providing a different viewpoint on some of the topics covered in Volume IV.

4. E. S. Yang, *Fundamentals of Semiconductor Devices*. New York: McGraw–Hill, 1978. See Chapters 7 and 8. Yang provides a condensed coverage of the subject matter which is quite readable and informative.

Appendix

LIST OF SYMBOLS

a	half-width of the channel region in a J-FET
A_G	gate area
C	capacitance; MOS-C capacitance
C_G	MOSFET gate capacitance
C_{gd}	gate-to-drain capacitance in the high-frequency, small signal equivalent circuit for the MOSFET
C_{gs}	gate-to-source capacitance in the high-frequency, small signal equivalent circuit for the MOSFET
C_o	oxide capacitance per unit area (pf/cm^2)
C_O	$C_O = C_o A_G$; oxide capacitance (pf)
C_S	semiconductor capacitance
D	drain
D_{IT}	density of interfacial traps (states/cm^2-eV)
D_N	electron diffusion constant
D_{ox}	dielectric displacement in the oxide
D_{semi}	dielectric displacement in the semiconductor
$\mathscr{E}$, $\mathscr{E}$	electric field
$\mathscr{E}_{ox}$	electric field in the oxide
$\mathscr{E}_S$	surface electric field; electric field in the semiconductor at the oxide–semiconductor interface
$\mathscr{E}_y$	y-direction component of the electric field
$\mathscr{E}_{vac}$	electric field in a vacuum
E_c	lowest possible conduction band energy
E_F	Fermi energy or Fermi level

E_i	intrinsic Fermi level
E_v	highest possible valence band energy
E_{vacuum}	vacuum level, minimum energy an electron must possess to completely free itself from a material
f	frequency (Hz)
f_{max}	maximum operational frequency of a MOSFET, cutoff frequency
$F(U, U_F)$	field function [see Eq. (2.21)]
G	gate
g_d	drain or channel conductance
g_m	transconductance or mutual conductance
i_d	small signal drain current
I_D	dc drain current in a J-FET or MOSFET
I_{D0}	$V_G = 0$ saturation drain current in a J-FET
I_{Dsat}	saturation drain current
j	$\sqrt{-1}$
$\mathbf{J}_N$, J_N	electron current density
J_{Ny}	y-direction component of the electron current density
k	Boltzmann constant (8.62×10^{-5}eV/°K)
K_O	oxide dielectric constant
K_S	semiconductor dielectric constant
L	length of the J-FET or MOSFET channel
L_D	intrinsic Debye length
L_D^*	extrinsic Deybe length
n	electron carrier concentration (number of electrons/cm^3)
N_A	total number of acceptor atoms or sites/cm^3
n_{bulk}	electron concentration in the semiconductor bulk
N_D	total number of donor atoms or sites/cm^3
n_i	intrinsic carrier concentration
N_I	number of implanted ions/cm^2
p	hole concentration (number of holes/cm^3)
p_{bulk}	hole concentration in the semiconductor bulk
p_s	hole concentration at the semiconductor surface (number/cm^3)
q	magnitude of the electronic charge (1.60×10^{-19} coul)
Q_F	fixed oxide charge per unit area at the oxide–semiconductor interface
q_g	ac component of the charge/cm^2 on the gate
Q_G	dc gate charge per unit area
Q_I	implant-related charge/cm^2 located at the oxide–semiconductor interface

Q_{IT}	net charge per unit area associated with the interfacial traps
Q_M	total mobile ion charge per unit gate area in the oxide
Q_N	total electronic charge/cm^2 in the MOSFET channel
Q_{O-S}	charge per unit area located at the oxide–semiconductor interface
Q_S	total charge in the semiconductor per unit area of the gate
R_D	channel-to-drain resistance in a J-FET
R_S	source-to-channel resistance in a J-FET
S	source
T	temperature
U	electrostatic potential normalized to kT/q
U_F	semiconductor doping parameter
U_S	normalized surface potential, U evaluated at the oxide-semiconductor interface
$\hat{U}_S$	sign ($\pm$) of U_S
V	voltage, electrostatic potential
V_A	voltage applied across a *p-n* junction
V_{bi}	"built-in" *p-n* junction voltage
V_{BS}	back-to-source voltage
v_d	ac drain voltage
V_D	dc drain voltage
V_{DS}	drain-to-source voltage
V_{Dsat}	saturation drain voltage
V_{FB}	flat band voltage
v_g	ac gate voltage
v_g'	ac gate voltage applied to an ideal device
V_G	dc gate voltage
V_G'	dc gate voltage applied to an ideal device
V_{GB}'	gate-to-back voltage being applied to an ideal device
V_{GS}	gate-to-source voltage
V_{GS}'	gate-to-source voltage being applied to an ideal device
V_{ox}	electrostatic potential in the oxide
V_P	pinch-off gate voltage in a J-FET
V_T	inversion–depletion transition point gate voltage; MOSFET threshold or turn-on voltage
V_T'	ideal device inversion–depletion transition point gate voltage
V_W	defined voltage [see Eq. (5.22)]
V_δ	defined voltage [see Eq. (3.13)]
W	depletion width

W_{eff}	effective depletion width
W_T	depletion width when the semiconductor is biased at the inversion–depletion transition point
x_c	depth of the MOSFET channel
x_o	oxide thickness
x'_o	$x'_o \equiv K_S x_o / K_O$
Δ	frequency parameter in the exact charge C–V theory [see Eq. (3.23)]
ΔL	decrease in the channel length under post-pinch-off conditions
ΔQ	general designation for a change in charge
ΔQ_G	small change in the static gate charge/cm^2 corresponding to a small change in gate voltage
ΔV_G	difference between the actual device and ideal device gate voltage required to achieve a given semiconductor surface potential
ΔV_{ox}	voltage drop across the oxide
ΔV_{semi}	voltage drop across the semiconductor
γ_M	normalized centroid of mobile ion charge in the oxide
ε	semiconductor permittivity ($\varepsilon = K_S \varepsilon_0$)
ε_0	permittivity of free space (8.85×10^{-14} farad/cm)
μ_n	electron mobility
$\bar{\mu}_n$	effective electron mobility
$\bar{\mu}_p$	effective hole mobility
ρ	charge density (coul/cm^3)
ρ_{ox}	ionic charge density in the oxide
ϕ_F	$\phi_F = (kT/q)U_F$
Φ_M	metal workfunction
Φ'_M	$\Phi'_M = \Phi_M - \chi_i$; effective metal workfunction in an MOS structure
ϕ_{MS}	metal–semiconductor workfunction difference
χ	semiconductor electron affinity
χ'	$\chi' = \chi - \chi_i$; effective semiconductor electron affinity in an MOS structure
χ_i	insulator (oxide) electron affinity
χ_{Si}	silicon electron affinity
ω	radian frequency

Index

Acceptor-like, interfacial traps, 74
Accumulation, 25, 26, 27, 28, 39
Alkali ions, *see* Sodium
Aluminum (Al):
  energy band diagram, 60
  gate material, 21, 59, 96
  and postmetallization annealing, 75
  workfunction difference, 63
Annealing, 70, 71, 74–76
Ar annealing, 70

Back biasing, 97–98
Back contact, 22, 82, 97, 98
Barrier heights, *see* Electron affinity, Metal workfunction
Bias-temperature stressing, 64, 66, 67
Block charge diagrams, 25
  ac probed MOS-C, 46
  deeply depleted MOS-C, 55
  static biasing states, 26, 27
Body effect, *see* Back biasing
Boron, 21, 96
Built-in charge, *see* Fixed oxide charge
Bulk, 22, 28, 29
Bulk charge, 91

Capacitance:
  oxide, 45, 46
  semiconductor, 46
  See also Capacitance–voltage (C–V) characteristics
Capacitance–voltage (C–V) characteristics, 43–58
  deep depletion, 54–56
  delta-depletion analysis, 48–49
  doping dependence, 51, 52
  exact charge analysis, 50–53
  experimental, 45, 55
  high frequency, 43, 45, 47, 51, 54
  low frequency, 45, 47, 50–53
  measurement of, 43–44, 53, 54
  oxide thickness dependence, 52, 53
  qualitative theory, 43–47
  quasi-static measurement technique, 53
  temperature dependence, 53
  theoretical, 52–53
Channel, 5, 82
Channel conductance, *see* Drain conductance
Chlorine (Cl_2), 68
Chlorine neutralization, 68
Chrome (Cr), 75
Copper (Cu), workfunction difference, 63
Cutoff frequency, 100

Dangling bonds, 74, 75

Debye length:
extrinsic, 30
intrinsic, 30, 31
Deep depletion, 54–56
Degeneracy, 30
Delta-depletion approximation, 34, 35
Delta-depletion formulation, 32–36
Delta-function, 35, 48, 69
Depletion, 26, 27, 28, 39
Depletion mode MOSFET, 94
Depletion width, 35
and deep depletion, 54
effective, 51
equilibrium maximum, 36
Dielectric constant, 38
Dielectric displacement, 37, 69
Donor-like, interfacial traps, 72, 74
Doping parameter:
definition, 28, 29
determination procedure, 58
properties, 29
Drain, 4, 81
Drain conductance:
J-FET, 16, 17
MOSFET, 98–102

Effective mobility, 85–88, 101, 102
mathematical expression, 86
measurement of, 101, 102
position dependence, 86
voltage dependencies, 86, 87, 88
Electric field, 31, 32, 35, 37
Electron affinity, 22, 23, 60
Electrostatic potential:
normalized, definition, 28, 29
normalized, properties, 29, 30
surface, 29, 30
Energy band diagram, 22–25, 59–62
isolated insulator, 23
isolated metal, 22, 23
isolated semiconductor, 22, 23
real MOS structures, 59–62
Enhancement mode MOSFET, 94
Equivalent circuit:
J-FET, 17
MOS-C, 46
MOSFET, 99
Extrinsic Debye length, 30
Fermi level, 24
Field effect, 3
Field-oxide, 95
Fixed oxide charge, 69–70, 71
Flat band, 27, 28, 39, 94
Flat band voltage, 94

Gate, 4, 22, 81
Gate-oxide, 95
Gate voltage relationship, ideal device, 38
Gauss's law, 50
Generation, 47, 53, 54
Gettering, 67
Gold (Au), workfunction difference, 63
Gradual channel approximation, 12, 13, 92

HCl, 68
High frequency, *see* C–V characteristics
Hydrogen (H_2), 74, 75, 76

IGFET, *see* MOSFET
Instability, 64, 65, 67, 92
Insulator:
energy band diagram, 23
perfect, 22
See also Silicon dioxide
Interfacial traps, 70–77, 92
acceptor-like, 74
definition, 72
density of, 74, 76
donor-like, 72, 74
effect on C–V characteristics, 73–74
effect on MOSFET characteristics, 92
electrical model, 72, 73
physical model, 74, 75
Intrinsic Debye length, 30, 31
Inversion, 26, 27, 28, 39
Inversion layer, 34, 82
Ion implantation, 96
Ionized impurity scattering, 85

Junction field effect transistor (J-FET), 3–17
ac response, 16–17

basic device structure, 11
channel, 5
channel narrowing, 6, 7, 8
drain conductance, 16, 17
equivalent circuit, 17
modern device structure, 4
pinch-off, 6, 7, 8, 9
pinch-off gate voltage, 10
post-pinch-off, 9, 13, 17
qualitative theory of operation, 5–10
quantitative $I_D - V_D$ relationships, 10–15
saturation, 8, 9, 10, 13, 14, 15
small signal equivalent circuit, 17
small signal parameters, 17
square-law relationship, 15
transconductance, 16, 17
waterfall analogy, 9

Lattice scattering, 85
Light-tight box, 43
Lithium (Li), alkali ion, 67
Low frequency, *see* C–V characteristics

Magnesium (Mg), workfunction difference, 63
Metal-insulator-semiconductor (MIS), 21, 77
Metal-oxide-semiconductor capacitor, *see* MOS-C
Metal-oxide-semiconductor field effect transistor, *see* MOSFET
Metal-semiconductor workfunction difference, 59–64
Metal workfunction, 22, 23, 60, 61
Minority carrier lifetime, 53, 68
MIS structures, 21, 77
Mobile ions, 64–68, 77, 92
effect on MOSFET characteristics, 92
model, 67
normalized charge centroid, 77
MOS-C:
C–V characteristics, 43–56
energy band diagram, 24, 60
ideal, 22
physical device, 21, 23
MOSFET, 81–103
ac response, 98–102
action of drain bias, 82, 83, 84
action of gate bias, 82
basic structure, 81, 82
bulk-charge theory, 90–92
channel, 82
cutoff frequency, 100
C–V characteristic, 101, 102
depletion mode, 94
effect of nonidealities, 92
enhancement mode, 94
equivalent circuit, 99
gate-to-drain capacitance, 99, 100
pinch-off, 83, 84
post-pinch-off, 83, 84, 90, 91, 99
qualitative theory of operation, 81–85
saturation, 84, 90, 91, 92
short channel effects, 84, 103
small signal characteristics, 100–102
small signal equivalent circuit, 99
small signal parameters, 99
square-law theory, 89–90, 92
threshold voltage, 85, 92–98
MOST, *see* MOSFET
Mutual conductance, *see* Transconductance

N_2 annealing, 70, 71
Neutron activation, 67
NMOS, 94, 95

Ohmic contact, 22
Overlap capacitance, *see* MOSFET, gate-to-drain capacitance
Oxidation triangle, 71
Oxide growth, 70
Oxide thickness, determination procedure, 58

Phosphorus, 21, 67, 68, 96
Phosphorus stabilization, 67, 68
Phosphosilicate glass, 67, 68
Pinch-off, 6, 7, 8, 9, 83, 84
Pinch-off gate voltage, 10
PMOS, 94, 95
Poisson equation, 31
Polycrystalline silicon, 21, 96
Polysilicon gate, 21, 96
Postmetallization annealing, 75, 76
Potassium (K), alkali ion, 67

Potential, *see* Electrostatic potential
Probing station, 44
Pseudo-channel, 94, 95

Quasi-static, definition, 16
Quasi-static technique, 53

Saturation, *see* J-FET, saturation, MOSFET, saturation
Self-aligned gate, 100
Silicon dioxide (SiO_2):
  energy band diagram, 60
  growth of and the fixed charge, 70
  insulator in MOS structures, 21, 23, 59, 82
  mobile ion contamination, 67
  Si–SiO_2 interfacial traps, 72, 73, 75, 76
  stabilization of, 67, 68
Silver (Ag), workfunction difference, 63
Sodium (Na), 65, 67, 68, 92
Source, 4, 81
Square-law relationship:
  J-FET, 15
  MOSFET, 90
Stacking faults, 68
Substrate-bias effect, *see* Back biasing
Surface orientation, 70, 74, 96
Surface potential, 29, 30
Surface scattering, 86, 87
Surface states, *see* Interfacial traps

Threshold voltage:
  adjustment, 95–98
  definition, 85
  measurement of, 100
  relationships, 93–94
Tin (Sn), workfunction difference, 63
Transconductance:
  J-FET, 16, 17
  MOSFET, 98, 99
Trichloroethane, 68
Trichloroethylene, 68
Turn-on voltage, *see* Threshold voltage

Unipolar transistor, 4

Vacuum level, 22, 59, 60, 61

Water vapor, 75, 76
Workfunction, 22, 23, 59–64